Future of Cities in Asia

Rosilawati Zainol · Norzailawati Mohd Noor ·
Weng Marc Lim · Huey Fen Cheong
Editors

Future of Cities in Asia

Future Direction of Urban Planning, Technology and Environmental Sustainability

Springer

Editors
Rosilawati Zainol
Universiti Malaya
Kuala Lumpur, Malaysia

Weng Marc Lim
Sunway Business School
Sunway University
Selangor, Malaysia

Norzailawati Mohd Noor
International Islamic
University of Malaysia
Kuala Lumpur, Malaysia

Huey Fen Cheong
Universiti Malaya
Kuala Lumpur, Malaysia

ISBN 978-3-032-01015-5 ISBN 978-3-032-01016-2 (eBook)
https://doi.org/10.1007/978-3-032-01016-2

This Springer imprint is published by the registered company Springer Nature Switzerland AG
The registered company address is: Gewerbestrasse 11, 6330 Cham, Switzerland

Titles on the *Futures of Asia*

Future of Business in Asia: Exploring Pathways to Tomorrow's Business Landscape
Future of Cities in Asia: Future Direction of Urban Planning, Technology and Environmental Sustainability
Future of Economy in Asia: Geopolitical Diplomacy, Technological Revolutions and Sustainable Pathways
Future of Education in Asia: Sustainability, Technology and Internationalisation
Future of Healthcare in Asia: Advancing Equity, Ethics and Technology in Asian Healthcare
Future of Media in Asia: Artificial Intelligence, Digital Technology and Media Practice
Future of Work in Asia: Transforming Work Models, Workforce Dynamics and Workplace Technology

Acknowledgements

The editors thank all of those who contributed to the completion of this book and the *Futures of Asia* series that comprises seven volumes, including this one.

We extend our deepest appreciation to the chapter authors for their hard work and kind cooperation throughout the process. We are also grateful to the individuals and organisations that helped disseminate the calls for papers (CFPs) like the Regional Studies Association (RSA), various academic research groups on social media (particularly Facebook), and our respective institutions, including our faculties and international relations offices. Special thanks goes to our assistant, Fatin Najihah, for her invaluable support in the early stages of this project.

We are also deeply grateful for the constructive feedback provided by the peer reviewers of this book, as well as the opportunities, guidance and support from Springer's commissioning editor, Doris Bleier.

Additionally, we also acknowledge and thank the following scholars for their insightful feedback on individual chapter manuscripts, both published and unpublished in this volume:

Reviewed 2 Manuscripts

1. Evalina Zuraidi (Universitas Syiah Kuala, Indonesia)
2. Fahmida Nusrat (Primeasia University, Bangladesh)
3. Roheen Qamar (QUEST Nawabshah, Pakistan)
4. Syafiqah Nazurah Mukhtar (Universiti Malaya, Malaysia)

Reviewed 1 Manuscript

1. Aliyah Nur Zafirah Sanusi (International Islamic University Malaysia, Malaysia)
2. Aslı Bozdağ (Niğde Ömer Halisdemir University, Türkiye)
3. Choon Wah Yuen (Universiti Malaya, Malaysia)
4. Fatih Eren (Istanbul Technical University, Türkiye)
5. Fatma Zohra Hafsi (University of Guelma, Algeria)
6. Hooi Ling Khoo (Universiti Tunku Abdul Rahman, Malaysia)
7. Kapil Kumar Gavsker (Banaras Hindu University, India)
8. Laina Hilma Sari (Universitas Syiah Kuala, Indonesia)
9. M. Zainora Asmawi (International Islamic University Malaysia, Malaysia)
10. Mila Karmilah (Universitas Islam Sultan Agung, Indonesia)
11. Noor Suzilawati Rabe (International Islamic University Malaysia, Malaysia)
12. Sami Zerari (University of Mohamed Khider Biskra, Algeria; University of Campania "Luigi Vanvitelli", Italy)
13. Sayyed Nouruddin Amiri (Persian Gulf University, Iran)
14. Sharaniya Vijitharan (University of Vavuniya, Sri Lanka)
15. Syakir Amir Abdul Rahman (International Islamic University Malaysia, Malaysia)

We also thank peer reviewers who opted for anonymity.

Again, we extend our deepest gratitude for the support, time, and expertise of these individuals. Their contributions were instrumental in publishing this book. In case we have unintentionally overlooked any individual or organisation that contributed to this project, we express our sincere appreciation and gratitude as well.

Contents

List of Figures

List of Tables

About the Editors

Rosilawati Zainol is a professor in urban analytics at the Department of Urban and Regional Planning, Faculty of Built Environment, Universiti Malaya. She currently holds the position of Deputy Dean for Undergraduate Studies for the faculty. She specialises in urban analytics, spatial analysis, sustainable tourism planning, and walkability.

Her research interests focus on the use of information technology in urban and regional planning, particularly GIS and spatial statistics. She has published numerous articles, book chapters, and books on sustainable urban and tourism development. Some of her recent academic works include studies on sustainable tourism governance, brownfield sustainability, child-friendly city assessment tools, and sustainable transportation systems.

Norzailawati Mohd Noor is a professor and deputy dean at the Kulliyah of Architecture and Environmental Design, International Islamic University Malaysia, Malaysia. Her expertise lies in urban planning technologies, particularly Geographic Information Systems (GIS) and remote sensing. With a Ph.D. and M.Sc. from UTM, her research focuses on integrating advanced geospatial tools into urban planning to address challenges like urban sprawl, land use change, and disaster resilience. As Vice President of IGRSM and a registered Town Planner, she promotes the use of spatial technologies in policy-making and planning practices. Her work bridges academic research, technological innovation, and practical implementation in sustainable urban development.

Weng Marc Lim is a Distinguished Professor and the Dean of Sunway Business School at Sunway University, an ASU-Cintana Alliance Global Partner Affiliate Faculty at Arizona State University in the USA, and an Adjunct Professor at Swinburne University of Technology in Australia and Sungkyunkwan University in the Republic of Korea. Professor Lim is the Record Holder for "Youngest Professor" in the Malaysia Book of Records and features in the Academy of Sciences Malaysia's Top Research Scientists Malaysia and JCI's Ten Outstanding Young Malaysians. With over 100 publications in prestigious "A*" and "A" ranked journals, Prof. Lim's research has been cited more than 45,000 times. As a thought leader, Prof. Lim has led high-level discussions at AppliedHE, Association to Advance Collegiate Schools of Business (AACSB), Association of MBAs (AMBA), Business Graduates Association (BGA), Quacquarelli Symonds (QS), Times Higher Education (THE), United Nations Educational, Scientific and Cultural Organization (UNESCO) and Principles of Responsible Management Education (UNPRME), Nikkei Forum and World Economic Forum (WEF). Professor Lim holds a Bachelor of Business and Commerce with First Class Honors and a Doctor of Philosophy from Monash University, and multiple certificates post-Ph.D., including a Graduate Certificate in Learning and Teaching (Higher Education) with High Distinction from Swinburne University of Technology and a Certificate in School Management and Leadership from Harvard University.

Huey Fen Cheong is a Senior Lecturer at Universiti Malaya, Malaysia. Her interests have been multidisciplinary, mainly linguistics, sociology, education, marketing, and psychology. Driven by a cultural and postcolonial perspective, she collaborated with a group of established scholars to initiate this *Futures of Asia* series (which contains this book) and the earlier *COVID-19 in Malaysia* series. These initiatives aim to promote Global South research and publications, especially on significant events such as the COVID-19 pandemic, post-pandemic normality and transformations, and the rise of AI, which equally matter to non-Western regions and deserve equal scholarly attention (and acceptance) beyond Western-centric frameworks in academia.

Editors' Note

The clusters of future studies in each chapter are based on Fergnani (2019, 2020), who identified six (6) key research/thematic clusters from reviews of future studies since 1968: "corporate foresight; past and futures; humanity at the limen; environmental futures; post-normality and complexity, and technological trends". While the chapter topics present different aspects of cities in Asia, the clusters of future studies show different dimensions of futures and future studies.

References

Fergnani, A. (2019). Mapping futures studies scholarship from 1968 to present: A bibliometric review of thematic clusters, research trends, and research gaps. *Futures*, *105*: 104–23.

Fergnani, A. (2020). Mapping fifty years of futures studies scholarship (1968–2017). In R. Slaughter & A. Hines (Eds.), *The Knowledge Base of Futures Studies 2020* (pp. 24–47). US: Association of Professional Futurists and Foresight International.

Chapter 1
Introduction: Framing the Future of Asian Cities

Rosilawati Zainol and Norzailawati Mohd Noor

Abstract This introductory chapter, *Introduction: Framing the Future of Asian Cities*, provides an overview of the central themes and discussions featured in *Future of Cities in Asia*. It delves into the ongoing urban transformations across Asia, driven by rapid urbanisation, technological progress, and the increasing effects of climate change. The chapter explains how urban planning, governance, and sustainability adapt to these developments. It reviews each chapter, shedding light on crucial topics like the lack of planning, the emergence of smart cities, environmental resilience in coastal areas, and the changing nature of workspaces within urban environments. This chapter lays the groundwork for detailed examinations of how Asian cities address the multifaceted challenges of modern urbanisation while seizing opportunities for sustainable, inclusive, and resilient growth. It advocates for comprehensive strategies that combine technological innovation with social equality and environmental sustainability to shape the future of urban life.

Keywords Future cities · Urban governance · Technological innovations · Inclusive urban growth · Sustainable development

1.1 Introduction

Asia is home to some of the fastest-growing cities in the world, with urban populations expanding at unprecedented rates (UN-Habitat, 2020). These cities are the beating heart of a global urban revolution. Home to 2.4 billion urban residents in 2023—over half the world's city dwellers. Asia is projected to house 3.5 billion by 2050, driven by

R. Zainol (✉)
Department of Urban and Regional Planning, Faculty of Built Environment, Universiti Malaya, Kuala Lumpur, Malaysia
e-mail: rosilawatizai@um.edu.my

N. M. Noor
Department of Urban and Regional Planning, Kulliyyah of Architecture and Environmental Design, International Islamic University Malaysia, Kuala Lumpur, Malaysia
e-mail: norzailawati@iium.edu.my

R. Zainol et al. (eds.), *Future of Cities in Asia*,
https://doi.org/10.1007/978-3-032-01016-2_1

migration, economic growth, and demographic shifts (United Nations, 2018). This rapid urbanisation, while driving economic growth and technological innovation, has brought about unique challenges, including environmental degradation, socio-economic inequality, and urban governance complexities. Addressing these issues requires a forward-looking approach that anticipates future needs while learning from the past. Future trajectories of the cities must be explored (Kundu & Pandey, 2020). Iksan et al. (2023) agree that understanding the trends is crucial in addressing these issues.

Future studies in urbanism provide a framework for envisioning how Asian cities can evolve sustainably. Pomeroy and Mookherjee (2023) highlight the emergence of interconnections among various discourses in urban transformations. Unlike traditional urban planning, which often reacts to immediate demands, future-oriented research focuses on long-term strategies that balance growth with resilience and inclusivity (Fergnani, 2019). By analysing trends and exploring scenarios, future studies empower stakeholders to make informed decisions that align with global sustainability goals, such as those outlined in the United Nations Sustainable Development Agenda (United Nations, 2015).

Urbanisation in Asia is not a homogeneous process. Cities within the region differ significantly in their cultural, political, and economic contexts. For instance, megacities like Tokyo and Jakarta face distinct challenges compared to smaller, rapidly growing cities in Indonesia or secondary cities in Vietnam. This diversity underscores the need for tailored solutions that respect local contexts while addressing global challenges. Likewise, Liu (2021) also raise the importance of considering the local context in utilising emerging technologies in urban planning.

Moreover, the future of Asian cities cannot be considered in isolation from broader global issues. Climate change, technological advancements, and geopolitical dynamics reshape the urban landscape (UN Habitat, 2021). As such, there is an urgent need to explore how these external factors influence Asian cities and how Asian urban innovations can contribute to global sustainability efforts.

This book examines these dynamics through the lens of future studies to shed light on innovative pathways for urban evolution in Asia. Integrating past and present insights into future planning frameworks ensures that urban development not only meets the needs of today but also anticipates and addresses the uncertainties of tomorrow.

1.2 The Meaning of Cities

Cities are dynamic hubs of human activity, shaping and shaped by culture, economics, politics, and environment. They are more than mere collections of buildings; they represent complex systems where human aspirations, innovations, and challenges converge. The meaning of cities varies across perspectives: geographic, sociological, economic, and cultural, each offering unique insights into their role in human civilisation. Below, we examine the significance of cities and the diverse perspectives that shape them, drawing on scholarly sources.

Cities are often described as engines of progress, fostering innovation, economic growth, and cultural exchange. They concentrate resources, ideas, and people, enabling advancements in technology, governance, and art (Glaeser, 2011). Urban areas facilitate specialisation and division of labour, driving economic productivity through economies of scale (Scott & Storper, 2015). Beyond economics, cities are cultural crucibles, where diverse populations interact, creating vibrant social fabrics and hybrid identities (Sassen, 2005). However, cities also embody contradictions: they are sites of opportunity and inequality, connectivity and alienation, sustainability and environmental strain.

From a historical perspective, cities have been central to civilisation since ancient times, from Mesopotamia's Uruk to modern megacities like Tokyo. They serve as nodes of power, trade, and knowledge, often reflecting the values and priorities of their societies (Mumford, 1961). Today, with over half of the global population living in urban areas, a figure projected to reach 68% by 2050, cities are pivotal in addressing global challenges like climate change, inequality, and technological disruption (United Nations, 2018).

1.3 The Perspectives of Cities

Cities can be viewed from four main perspectives, namely, geographic, sociological, economic and cultural. Geographers view cities as spatial systems shaped by their location, connectivity, and the distribution of resources. Urban morphology refers to the physical structure of cities, reflecting historical and environmental factors, such as the presence of rivers or trade routes (Pacione, 2009). For example, coastal cities like New York thrive due to port access, while inland cities like Brasília were planned to catalyse regional development. Geographers also study urban sprawl and its environmental impacts, emphasising sustainable urban planning to balance growth and ecological limits (Bruegmann, 2005).

Sociologists focus on cities as social ecosystems. Cities amplify human interaction, fostering both community and anonymity. Robert Park's urban ecology theory likens cities to organisms, with neighbourhoods competing and adapting within a larger system (Park, 1915). Social stratification is a key concern: cities often segregate by class, race, or ethnicity, creating enclaves of privilege and poverty (Sampson, 2012). Yet, they also enable social mobility and cultural diversity, as seen in global cities like London or Shanghai, where multiculturalism drives innovation but also social tension (Sassen, 2005).

Economists highlight cities as engines of economic growth and wealth creation. The clustering of people and businesses in close proximity enhances productivity, making cities appealing hubs for enterprises and employees (Glaeser, 2011). However, urban economies also face challenges like gentrification, where rising costs displace lower-income residents (Lees et al., 2016). The rise of "smart cities," integrating technology for efficiency, reflects economic priorities but raises concerns about surveillance and inequity (Townsend, 2013).

Culturally, cities are stages for identity and expression. They host festivals, art scenes, and subcultures, shaping global cultural trends (Zukin, 1995). Cities like Paris or Mumbai are not just places but symbols of romance, creativity, or resilience. However, globalisation can homogenise urban cultures, as seen in the spread of chain stores, eroding local distinctiveness (Sassen, 2005).

Cities face pressing challenges: climate change demands resilient infrastructure, while inequality requires inclusive policies. The rise of digital technologies presents opportunities for more innovative urban systems but also risks exacerbating existing divides (Townsend, 2013). Future cities must strike a balance between growth and sustainability, diversity and cohesion, and innovation and equity.

Cities are multifaceted entities, embodying human potential and paradoxes. Their meaning evolves in response to societal needs, shaped by geographic, sociological, economic, and cultural factors. As urbanisation accelerates, understanding these perspectives is crucial for building cities that are liveable, equitable, and sustainable.

1.4 Book Structure and Overview

1.4.1 Part I: Future Direction of Urban Planning: Complexity and Governance

Urban planning in Asia is deeply influenced by the complexities of governance, socio-political dynamics, and the rapid pace of urbanisation. With cities in the region experiencing unprecedented growth, managing urban spaces has become challenging. The interplay of formal and informal systems, historical legacies, and emerging governance models shapes the region's **Future Direction of Urban Planning: Complexity and Governance**.

Governance systems in Asian cities must address the multifaceted nature of urban environments, characterised by diverse populations, resource constraints, and overlapping jurisdictions. Many cities, such as Jakarta and Manila, face significant challenges in coordinating local, regional, and national policies (UN-Habitat, 2020). Governance frameworks must evolve to manage these complexities effectively, ensuring that urban areas can accommodate growth while addressing key issues such as housing, transportation, and essential services. Integrating urban management systems and fostering regional collaboration is essential for addressing governance fragmentation and achieving sustainable outcomes (ASEAN, 2020).

Chapter 2 reveals how historical city planning has profoundly shaped the urban fabric of Asian cities, such as Manila, where colonial-era governance, grid-based spatial layouts, and infrastructure decisions continue to influence modern urban development, land use patterns, and social dynamics (Hammes, 2023; King, 2004;

Meller, 2018). Informal urbanisation, a hallmark of cities such as Mumbai, Manila, and Jakarta, emerges from inadequate planning, acute housing shortages, and deep socioeconomic disparities, with slum areas housing large populations outside formal urban frameworks (Roy, 2009). These settlements demonstrate remarkable community resilience, adaptive ingenuity, and vibrant local economies, yet they also highlight the limitations of conventional urban planning in addressing rapid urban growth. Questions are also raised about whether the planning systems employed can improve liveability. Thus, to foster inclusivity, enhance access to essential services like water, sanitation, and electricity, and promote sustainable urban development, future planning strategies must integrate these informal spaces into broader city systems, ensuring equitable growth and resilience (UNESCAP, 2019).

Chapter 3 explores the concept of urban amnesia, defined as the erosion of a city's historical and cultural memory due to rapid urban transformation, globalisation, migration, and extraordinary events like wars and disasters (Crinson, 2005; Rastouskaya, 2019). Urban amnesia leads to the erosion of a city's unique identity, disrupted cultural knowledge transfer, and weakened social cohesion. It is prevalent in historical cities (e.g., Jakarta, Beijing), industrial cities (e.g., Tokyo, Detroit), post-conflict cities (e.g., Kabul, Hiroshima), global hubs (e.g., Singapore, New York), former colonial cities (e.g., Mumbai, Hanoi), and multicultural metropolises (e.g., Dubai, Sydney).

Focusing on Gaza City, the chapter details eight key sites that embody urban amnesia due to the prolonged conflict since 1948, including the Great Omari Mosque, Gaza Port, the Traditional Gaza Marketplace, Gaza Palestine Square, the Islamic University of Gaza, Al-Shifa Hospital, Gaza Beach, and Mathaf al-Funduq. These sites, damaged by wars, blockades, and the 2023 occupation, reflect Gaza's resilience and memory loss, with 74.3% of buildings destroyed. Other Asian cities, such as Kabul, Lahore, Srinagar, Yangon, Bangkok, Hanoi, Seoul, Tehran, Beirut, Jakarta, and Bishkek, exhibit similar patterns of memory erosion due to war damage, rapid modernisation, and political upheaval.

In addressing this issue, planning strategies must prioritise resilience and adaptability in such contexts to support recovery and long-term development. Lessons from these cities can inform planning practices in other regions experiencing similar challenges, such as post-conflict reconstruction and disaster management (Watson, 2014).

Chapter 4 examines the development of satellite cities in Asia, focusing on Hung Yen City (HYC), Vietnam, as a case study compared with satellite cities in China, Japan, and India. Satellite cities, defined as smaller urban areas near larger metropolitan centres, serve as residential, economic, or administrative hubs to alleviate urban pressures, such as overpopulation and housing shortages.

The chapter emphasises the importance of inclusive, decentralised planning and diversified funding to ensure the sustainable development of satellite cities. It advocates for co-produced urban solutions to balance technological innovation with equitable growth, warning against neoliberal models that prioritise elites and risk creating ghost towns (Abubakar & Doan, 2017; Van Noorloos & Kloosterboer, 2018). HYC's case contributes to Asian urbanism by highlighting the complexities of authoritarian governance and South-South urban inter-referencing in shaping future cities (Bunnell et al., 2012).

1.4.2 Part II: Future Direction of Urban Technology: Innovation and Smart Cities

Technological innovation is at the forefront of urban transformation, particularly in rapidly urbanising regions such as Asia. The push towards adopting smart city frameworks underscores the critical role of technology in addressing urban challenges, including congestion, pollution, inefficient infrastructure, and resource management. The part, **Future Direction of Urban Technology: Innovation and Smart Cities**, explores the potential of digital innovations to reshape urban systems, governance, and the quality of life for residents.

Chapter 5 introduces a digital twin framework to address Bangkok's complex transportation challenges, characterised by severe traffic congestion, inadequate public transit coverage, and poor air quality. A digital twin, a virtual replica of physical systems using real-time data, is presented as a transformative tool for urban planning (Attaran & Celik, 2023). This framework offers significant potential to create a sustainable, efficient, and inclusive transportation system in Bangkok, contributing to broader discussions on smart urban solutions. The proposed framework for Bangkok, as described in the chapter, includes a data collection layer (utilising IoT sensors, GPS, and cameras), a data integration platform, a simulation engine, an AI-driven analytics module, and a visualisation interface. These components enable simulations of traffic flow, public transit optimisation, and air quality management, supporting data-driven decisions (Deng et al., 2021). The implementation process involves deploying sensors, standardising data, and building a 3D model of the transport network. However, challenges include data quality, privacy concerns, costs, and stakeholder resistance.

Chapter 6 reveals the application of smart city adoption in city planning. It describes how the motivation for smart city adoption can balance the tension between push and pull factors. It also describes the benefits gained from the application. It highlights the interplay of normative and latent motives, revealing the non-linear, politically driven nature of smart city policymaking. While smart city initiatives aim to address urban challenges and align with the Sustainable Development Goals, latent motives such as political legitimacy and city branding risk superficial development if not balanced with tangible outcomes. Future smart city planning should balance push and pull factors, integrate stakeholder interests, and prioritise human-centric, environmentally adaptive designs.

Chapter 7 revolves around the concept of future cities, which envisions sustainable, resilient, and inclusive urban environments to address rapid urbanisation, climate change, technological advancements, and socio-economic disparities. Smart city initiatives in places like Tokyo, Barcelona, and Dubai leverage IoT, AI, and big data for efficient urban management, offering models for Southeast Asian cities like Malaysia's Cyberjaya, Indonesia's Nusantara, and the Philippines' New Clark City to enhance infrastructure, transportation, and energy efficiency, while cities like Amsterdam and Vienna inspire human-centric planning to ensure service access for underserved communities. In South Asian cities, such as India's Amaravati and Colombo Port City, the focus is on smart technology. However, they face financial and institutional challenges. In contrast, East Asian cities like Songdo and Xiong'an lead with advanced IoT and green infrastructure but must address ageing populations and inclusivity. Future directions of smart cities include addressing these gaps through smart mobility, green infrastructure, and digital governance, combating climate risks with resilient systems, promoting sustainability via renewable energy and circular economies, and tackling disparities in cities like Manila and Jakarta with inclusive policies. Undoubtedly, political barriers, fragmented governance, and the need for USD210 billion annually in infrastructure investment pose significant challenges. At the same time, AI, IoT, and 5G promise enhanced urban management, but they also raise concerns about data privacy and digital equity. Thus, the balance between these challenges needs to be sought.

1.4.3 Part III: Future Direction of Environmental Sustainability: Adaptation and Resilience

Environmental sustainability is at the core of urban transformation in Asia, where cities face mounting pressures from climate change, resource scarcity, and rapid urbanisation. The part, **Future Direction of Environmental Sustainability: Adaptation and Resilience**, explores innovative pathways for cities to thrive amid environmental challenges, drawing on global insights to foster resilience and sustainability. It further explores strategies for reimagining urban spaces to address ecological challenges while fostering resilience and inclusivity. This perspective integrates adaptive planning, nature-based solutions, and innovative urban policies to ensure sustainable futures for Asian cities.

Asian cities are disproportionately vulnerable to climate change due to their geographical contexts and rapid urbanisation. Coastal cities like Jakarta, Bangkok, and Manila face rising sea levels, flooding, and saltwater intrusion, while inland cities like Delhi grapple with heat waves, water scarcity, and air pollution (UN Habitat, 2021). These challenges underscore the urgency of urban transformation in building resilience against environmental risks. Jakarta's relocation of its capital city to Kalimantan directly responds to chronic flooding and subsidence caused by the over-extraction of groundwater and rising sea levels (Syaban & Appiah-Opoku, 2023). Therefore, human creativity is expected to be the catalyst in addressing and balancing the use of natural resources to fulfil their economic goals. Additionally, institutional arrangements are likely to balance economic ambitions, challenges and environmental demand.

Chapter 8 reveals the challenges of city planning in facing environmental degradation along the coastline. A study conducted in Dukuh Timbulsloko, Demak Regency, Central Java, Indonesia, revealed the adaptive strategies employed by the community. They demonstrate social-ecological resilience through transformative adaptation, driven by economic constraints and deep attachment to their homeland, which discourages relocation. The findings of this study highlight the importance of integrating local knowledge with modern technologies and aligning community efforts with global frameworks, such as the Sustainable Development Goals, to ensure scalability and long-term resilience in Asian cities. Transformation requires not only overcoming vulnerability through awareness, knowledge, and action, but also government support, which is critical for effective planning and resource allocation. This chapter also highlights the other future Asian coastal cities that face similar challenges and strategies in overcoming them.

Chapter 9 explores the integration of natural resource conservation within the creative city model, focusing on UNESCO Creative Cities in the Philippines (Baguio, Cebu, Iloilo) and their Asian counterparts (Chiang Mai, Thailand; Singapore; Macao, China). The creative city model, which emphasises human creativity as a driver of sustainable development, often marginalises natural resources due to its economic orientation, risking cultural and environmental sustainability. The chapter argues that natural resources are foundational to cultural resources, which fuel creativity, and should be prioritised alongside human creativity for sustainable development. It further describes the future of nature-based solutions (NBS), which have emerged as transformative approaches to urban sustainability. By integrating natural systems into urban landscapes, cities can enhance biodiversity, improve air quality, and mitigate the effects of urban heat islands. Projects such as green roofs, urban forests, and rain gardens contribute to ecological balance while improving urban aesthetics (Elmqvist et al., 2018). Singapore's "City in Nature" initiative exemplifies the integration of urban greening into city planning. The Singapore Green Plan 2030 encompasses extensive green corridors, tree-planting initiatives, and urban farming projects aimed at enhancing liveability and environmental sustainability (Singapore Ministry of Sustainability and the Environment, 2021).

Chapter 10 explains the transformation of workspaces in response to the Fourth Industrial Revolution (4IR) and its impact on urban development in selected Asian cities: Shanghai (China), Seoul (South Korea), Bangkok and Chiang Mai (Thailand), Kuala Lumpur (Malaysia), and Dubai (United Arab Emirates). The shift from traditional cubicle-based offices to flexible, technology-driven workspaces reflects advancements in AI, robotics, and digital connectivity, which the COVID-19 pandemic has accelerated. Key concepts include flexible workspaces enhancing employee satisfaction, technological integration optimising work environments, cultural influences shaping work dynamics, and urban planning promoting sustainable development. The discussion highlights the importance of hybrid and activity-based workspaces, modular designs, biophilic elements, and smart technologies in the future of work environments. Implications include the need for urban planning to adapt to decentralised work patterns, address mental health, and promote equity.

Future directions emphasise integrating adaptable workspaces into urban frameworks to support economic growth, attract talent, and enhance resilience.

1.5 Clusters of Future Studies: Framing Asia's Urban Tomorrow

Asia's cities pulse with life, Manila's chaotic markets, Bangkok's neon-lit innovation, and Gaza City's quiet rebuilding amid scars. These urban landscapes are not just places but stories of ambition, struggle, and hope, each demanding a bold vision for the future. To navigate this complexity, these chapters turn to Alessandro Fergnani's groundbreaking work in futures studies, which maps the field's evolution through a bibliometric analysis of scholarship from 1968 to 2018 (Fergnani, 2019). Fergnani identifies six thematic clusters: corporate foresight, past and futures, humanity at the limen, environmental futures, post-normality and complexity, and technological trends, which offer a powerful framework for envisioning Asia's urban future. By applying these clusters to the diverse urban contexts explored in the book, ***Future of Cities in Asia***, from Vietnam's satellite cities to Java's climate battles, this book sets the stage for understanding how urban planning, technology, and sustainability can weave resilient, inclusive urban tomorrows.

Fergnani's clusters highlight the fragmentation of futures studies, with corporate foresight and past and futures dominating citations since 2004; yet, they offer a unified framework for urban Asia (Fergnani, 2019). By applying these clusters, this book envisions cities that integrate planning, technology, and sustainability to create resilient and equitable futures. From Jakarta's climate challenges to Dubai's connected workspaces, Fergnani's insights guide planners and communities toward collaborative, forward-thinking urban solutions.

Collectively, these six clusters enable urban stakeholders to anticipate and navigate complex challenges, from climate change and technological disruption to social inequities and historical legacies. By leveraging foresight methodologies like scenario planning, systems thinking, and participatory visioning, cities can move beyond reactive planning to proactively design sustainable, resilient, and inclusive urban futures. Each cluster emphasises human-centric, equitable, and adaptive strategies, ensuring that cities balance innovation, ecological health, and social cohesion in an era of uncertainty. Together, they offer a comprehensive roadmap for urban transformation, aligning local actions with global imperatives for liveable, future-ready cities (Fergnani, 2019).

1.5.1 Corporate Foresight

The future of cities is shaped by complex, interconnected challenges, including climate change, population growth, technological disruption, and social inequalities. The corporate foresight cluster provides a structured approach for urban stakeholders to navigate these challenges by anticipating future urban scenarios and aligning strategies accordingly. It also enables city planners and businesses to scan for weak signals, early indicators of change, and develop scenarios to prepare for disruptions. Fergnani (2022) emphasises that scenario planning helps stakeholders move beyond binary (positive/negative) views of the future, encouraging critical thinking and exploration of a spectrum of possibilities. In urban planning, this cluster can prioritise climate-related themes like resilience, sustainable infrastructure, and stakeholder collaboration, enabling focused strategies.

Corporate foresight in urban planning involves strategic anticipation and preparation for future challenges, particularly those driven by environmental and socio-economic shifts. Chapter 8, focusing on Timbulsloko, Indonesia, and other coastal Asian cities, aligns with this cluster by addressing the proactive strategies needed to combat the impacts of climate change, such as rising sea levels and coastal erosion. Urban planners in these regions must employ foresight to design resilient infrastructure, such as elevated buildings, flood barriers, and sustainable drainage systems, to mitigate flooding risks. The chapter likely explores how corporate entities, including governments and private developers, collaborate to integrate climate adaptation into urban development plans. For instance, Timbulsloko's case highlights the need for innovative engineering solutions, like floating structures or mangrove restoration, to preserve coastal ecosystems while supporting urban growth. This cluster emphasises data-driven scenario planning, where planners use predictive models to anticipate climate impacts and allocate resources effectively. By integrating corporate foresight, cities can transition from reactive disaster management to proactive resilience, ensuring long-term habitability and economic stability in vulnerable coastal zones.

1.5.2 Past and Futures

The past and futures cluster is a thematic area in futures studies that explores the historical trajectories of urban systems and uses these insights to anticipate and design future urban scenarios. Fergnani's (2019) analysis reveals that this cluster integrates retrospective analysis (examining how cities have developed over time) with prospective methodologies (such as scenario planning, visioning, and trend analysis) to address urban challenges like sustainability, mobility, governance, and social inclusion. The cluster emphasises the importance of understanding path dependencies, how past decisions, infrastructure, and cultural factors shape current and future urban possibilities, while also imagining transformative urban futures. This cluster provides a framework for urban stakeholders, planners, policymakers, businesses, and communities to learn from historical urban patterns and proactively shape future urban environments.

The "past and futures" cluster examines how historical urban planning decisions shape future trajectories. Chapter 2 explores Manila's urban fabric through a historical lens, likely analysing how unplanned, organic growth has led to challenges like congestion, informal settlements, and inadequate infrastructure. From an urban planning perspective, this chapter underscores the importance of learning from past non-planning to design structured, inclusive urban frameworks. For example, Manila's chaotic urban sprawl could inform policies for zoning, affordable housing, and public transport to prevent future inefficiencies.

Chapter 3, which focuses on Gaza City and conflict-prone Asian cities, explores how prolonged conflict leaves lasting spatial impacts, such as destroyed infrastructure or disrupted urban grids, that influence future planning. Urban planners must address "urban amnesia," where historical urban identities are erased, by preserving cultural landmarks and integrating peacebuilding into reconstruction efforts. This could involve participatory planning to rebuild trust and ensure equitable distribution of resources in post-conflict cities.

Chapter 5's focus on Bangkok's digital twin revolution highlights a forward-looking approach, where historical transport challenges (e.g., traffic congestion) inform futuristic solutions. Digital twins—virtual models of physical urban systems—enable planners to simulate traffic flows, optimise public transport, and reduce emissions, creating a sustainable transport landscape. This chapter likely discusses how past inefficiencies in Bangkok's transportation system have driven the adoption of advanced technologies.

Chapter 6, centred on Magelang City, Indonesia, examines smart city adoption, blending historical urban governance with futuristic technological integration. Smart cities rely on data analytics and IoT to enhance urban services, but political dynamics, such as stakeholder conflicts or resource allocation, shape their success. Urban planners must navigate these complexities to ensure smart city initiatives align with historical urban needs while fostering equitable access to technology.

1.5.3 Humanity in the Limen

Humanity at the limen, focusing on societal thresholds, captures the transitions reshaping Asian urban life (Fergnani, 2019). It explores the human and social dimensions of urban life during transitional phases, where cities navigate disruptive changes such as technological advancements, climate crises, or socio-economic upheavals. Fergnani's analysis positions this cluster as a critical lens for understanding how urban dwellers experience, adapt to, and shape these liminal moments. It draws on futures studies methodologies such as narrative foresight, causal layered analysis (CLA), and participatory visioning to examine how human values, identities, and communities evolve in response to urban transformation. Fergnani (2019) notes that this cluster is particularly relevant in an era of rapid urban change, where cities face existential challenges like climate change, automation, and social polarisation. It seeks to ensure that human well-being remains at the core of urban futures, even as cities undergo profound transformations.

The "humanity at the limen" cluster focuses on urban populations at critical transitional thresholds, grappling with social, political, and spatial disruptions that challenge their identity, resilience, and future development. Chapter 3, aligns with this cluster by examining the profound impacts of prolonged conflict on urban environments and their inhabitants, who exist at the limen, a threshold between destruction and recovery, memory and erasure, stability and uncertainty. From an urban planning perspective, this chapter underscores the complex role of planners in navigating the spatial, cultural, and social legacies of conflict to foster equitable and sustainable urban futures. Planners must prioritise participatory approaches to ensure marginalised voices shape urban recovery, addressing the social and psychological impacts of prolonged occupation.

Chapter 9 focuses on natural resource conservation in creative cities across the Philippines, Thailand, Singapore, and China. These cities are at a threshold where economic growth and environmental sustainability must coexist. Urban planners can promote green infrastructure, like urban forests or rainwater harvesting systems, to conserve resources while fostering creative industries. This chapter likely explores how cultural and economic vibrancy can align with sustainable urban planning to support populations navigating resource scarcity. Likewise, this cluster champions urban planning that navigates transitional moments with empathy, ensuring cities evolve as inclusive, human-centred spaces.

1.5.4 Environmental Future

The environmental future for cities cluster examines how cities can address environmental challenges and transition to sustainable, resilient urban systems. Fergnani's analysis highlights this cluster as a critical area of futures studies, driven by the urgent need to mitigate climate change, reduce urban ecological footprints, and adapt to environmental disruptions. The cluster integrates foresight methodologies such as scenario planning, backcasting, and environmental scanning with urban planning and environmental science to envision cities that balance human needs with ecological health. Cities are on the frontline of environmental challenges, from climate-induced flooding to air pollution and resource scarcity. The Environmental Future for Cities cluster provides a framework for urban stakeholders, planners, policymakers, businesses, and communities to anticipate these challenges and design cities that are ecologically sustainable and resilient.

The environmental futures cluster addresses how urban planning must respond to ecological challenges, and Chap. 2's analysis of Manila's non-planning aligns with this cluster by highlighting how historical urban development patterns have exacerbated environmental vulnerabilities, shaping the city's future sustainability. Colonial and postcolonial planning models, often rigid and disconnected from local realities, failed to address rapid population growth and economic pressures, leading to uncontrolled urban expansion. This non-planning has contributed to environmental degradation, such as flooding exacerbated by poor drainage systems, deforestation from urban encroachment, and pollution from unregulated settlements along rivers.

Smart city adoption, as discussed in Chap. 6, typically involves technologies like IoT, data analytics, and smart infrastructure to enhance urban efficiency, which often includes environmental sustainability. In Magelang, a city in Central Java with significant cultural and ecological importance, smart city initiatives could address environmental challenges such as waste management, energy efficiency, and disaster resilience, aligning with the Environmental Futures cluster.

Chapter 7 aligns closely with this cluster, embodying urban transformation through advanced technologies, sustainable practices, and innovative planning to tackle rapid urbanisation. It reflects the cluster's emphasis on sustainable, resilient, and equitable interactions between humans and their environment. Physical, social, and economic drivers such as smart sensors for data-driven urban systems, community-focused well-being, and green economic growth mirror the interdisciplinary approach of balancing ecological limits with human needs.

Chapter 8 focuses on Timbulsloko's coastal vulnerabilities, proposing nature-based solutions like mangrove restoration, coral reef rehabilitation, and elevated infrastructure to combat sea-level rise. The chapter likely explores how municipalities and planners in Timbulsloko and similar Asian coastal cities anticipate future risks and implement adaptive measures, such as infrastructure investments or community relocation plans, embodying the forward-looking essence of corporate foresight in urban governance.

Chapter 9 focuses on natural resource conservation in creative cities, emphasising the importance of sustainable urban design. Planners can implement policies such as zero-waste initiatives or green building standards to reduce ecological footprints while supporting local cultural economies. It also highlights the institutional arrangements of creative cities and reframes the creative city model as a tool for conserving natural resources.

Chapter 10 explores the future of workspaces in connected cities, where environmental sustainability shapes urban design. Planners must create energy-efficient office spaces, promote remote work to reduce commuting emissions, and integrate green spaces to enhance worker well-being. This chapter likely discusses how cities like Singapore and Dubai strike a balance between economic productivity and environmental goals.

1.5.5 Post-normality and Complexity

The Post-normality and Complexity cluster examines how complex systems, such as cities, operate under conditions of post-normality, where high uncertainty, contradictory values, and interconnected risks challenge conventional planning and decision-making. Post-normality, a concept rooted in the work of Funtowicz and Ravetz, describes situations where facts are uncertain, stakes are high, values are disputed, and decisions are urgent. Fergnani's analysis positions this cluster as a critical lens for futures studies, emphasising the need for non-linear foresight methodologies (e.g., scenario planning, systems thinking, and causal layered analysis) to address the emergent, unpredictable nature of complex urban systems.

Cities are quintessential complex systems, with millions of interacting components: people, infrastructure, ecosystems, and technologies operating in unpredictable, non-linear ways. The Post-normality and Complexity cluster provides a framework for urban stakeholders, planners, policymakers, businesses, and communities to navigate these dynamics and build resilient, adaptive cities. This cluster addresses the unpredictable, interconnected challenges of modern urban systems.

Chapter 8 examines the impacts of climate change, particularly rising sea levels, on Timbulsloko, Indonesia, and other Asian coastal cities, emphasising strategies for building resilient urban systems. Climate change is a quintessential post-normal challenge, involving interdependent variables such as rising sea levels, extreme weather events, economic constraints, and social displacement. For instance, resilience strategies in Timbulsloko might include elevating homes, building seawalls, or relocating communities, each option entangled with ecological, financial, and cultural trade-offs. These decisions are non-linear, as small changes (e.g., a single storm) can escalate impacts, and solutions must account for uncertainty, aligning with the cluster's emphasis on complex systems.

1.5.6 Technological Trends

Finally, the technological trends in the futures of cities cluster examine how emerging and disruptive technologies influence urban environments, including infrastructure, governance, mobility, and social dynamics. Fergnani's analysis highlights this cluster as a growing area of futures studies, driven by the rapid pace of technological advancement and its transformative potential for cities. The cluster employs foresight methodologies, including trend analysis, scenario planning, and horizon scanning, to identify technological drivers, explore their urban applications, and mitigate risks such as inequality and loss of human agency. Cities are hubs of technological innovation, but their adoption of new technologies introduces opportunities and risks. The technological trends cluster provides a framework for urban stakeholders, planners, policymakers, businesses, and communities to anticipate technological trends, harness their benefits, and mitigate downsides.

The technological trends cluster focuses on how emerging technologies shape urban planning. Chapter 4's exploration of satellite cities likely includes technological innovations, such as smart infrastructure or automated transport systems, to enhance urban efficiency. Satellite cities often leverage advanced infrastructure, such as smart grids or high-speed connectivity, to establish their brand identity and attract investment. Technologies such as smart infrastructure, digital governance, and energy-efficient systems support the branding and inclusivity goals of cities, enhancing their appeal and functionality. Vietnam's urban imaginary is likely to incorporate technology to compete with regional models, navigating complex adoption challenges.

Chapter 6's focus on Magelang's smart city adoption emphasises technologies like IoT and AI in urban governance. Smart cities are inherently tied to technological trends, relying on technologies like IoT, data analytics, and AI to optimise urban systems. In Magelang, smart city initiatives may include smart traffic management, energy-efficient infrastructure, or digital governance platforms, all of which directly align with emerging technological trends. The "politics of urban futures" likely involves debates over adopting these technologies, such as balancing cost, accessibility, and privacy.

Chapter 7 likely includes technological advancements as key drivers for Asian cities, such as 5G networks or renewable energy systems. Planners must strike a balance between technological adoption, affordability, and scalability. Technological trends are likely to manifest through innovations such as smart infrastructure, renewable energy systems, and digital platforms for urban governance. For instance, cities in the region may adopt AI-driven urban planning tools or 5G networks to enhance connectivity, thereby addressing challenges such as population growth and resource scarcity. The chapter's regional scope suggests it explores how technologies vary across contexts, e.g., Singapore's advanced smart city systems versus emerging tech adoption in less-developed cities. The explicit matrix linkage underscores technology's pivotal role in driving urban transformation, complementing environmental and systemic considerations.

Chapter 10's focus on connected workspaces highlights technologies like IoT-enabled offices or virtual collaboration platforms. Urban planners must design flexible, tech-enabled workspaces that support hybrid work models while promoting sustainability. The shift from cubicles to connected urban workspaces likely involves technologies like cloud computing, IoT, and smart building systems to enable hybrid work environments. Cities like Shanghai or Dubai might integrate digital platforms for remote collaboration or AI-driven office management. Nonetheless, the technological infrastructure underpinning connected cities implies an implicit link to this cluster, as workspaces increasingly rely on digital and smart technologies.

1.6 Conclusion: Towards a Shared Vision for Future Asian Cities

The challenges and opportunities facing Asian cities highlight the critical need for a holistic and inclusive approach to urban development. As rapid urbanisation reshapes the region, cities must balance growth demands with the imperatives of sustainability and equity. The rapid urbanisation of Asia presents both significant challenges and remarkable opportunities. Cities across the region are at the forefront of addressing complex issues, including climate change, technological advancements, and socioeconomic inequality. Addressing these complexities requires integrating diverse perspectives from policymakers, urban planners, local communities, and technology innovators.

Cities can craft contextually relevant and forward-thinking solutions by bringing together historical insights, technological advancements, and ecological strategies. Collaborative governance, participatory planning, and knowledge-sharing platforms, such as the ASEAN Smart Cities Network, exemplify the power of diverse perspectives in creating resilient and inclusive urban environments.

Most importantly, sustainable urban futures depend on recognising all stakeholders' unique needs and contributions, including marginalised groups often excluded from decision-making processes. This integration ensures that urban strategies are innovative and equitable, fostering a sense of ownership and shared responsibility among all city residents.

As Asian cities continue to evolve, their ability to draw on diverse perspectives will determine their success in building adaptable, inclusive, and environmentally sustainable communities. By reinforcing this collaborative vision, Asian cities can emerge as global leaders in sustainable urban development. This book aims to guide and inspire people to realise this ambitious but essential goal.

References

Abubakar, I. R., & Doan, P. L. (2017). Building new capital cities in Africa: Lessons for new satellite towns in developing countries. *African Studies, 76*(4), 546–565. https://doi.org/10.1080/00020184.2017.1370187

ASEAN. (2020). *ASEAN smart cities network annual report*. Retrieved November 7, 2024, from chrome-extension://efaidnbmnnnibpcajpcglclefindmkaj/https://www.miti.gov.my/miti/resources/ASEAN_Annual_Report_2020-2021.pdf

Attaran, M., & Celik, B. G. (2023). Digital twins: A review of concepts, technologies, and applications. *International Journal of Production Research, 61*(3), 1121–1137. https://doi.org/10.1080/00207543.2022.2094210

Bruegmann, R. (2005). *Sprawl: A compact history*. University of Chicago Press.

Bunnell, T., Goh, D. P. S., Lai, C.-K., & Pow, C.-P. (2012). Introduction: Global urban frontiers? Asian cities in theory, practice and imagination. *Urban Studies, 49*(13), 2785–2793. https://doi.org/10.1177/0042098012452455

Crinson, M. (2005). *Urban memory: History and amnesia in the modern city*. Routledge. https://doi.org/10.4324/9780203414613

Deng, T., Zhang, K., & Shen, Z. J. M. (2021). A systematic review of digital twins in urban planning and mobility. *Transport Reviews, 41*(6), 733–757. https://doi.org/10.1080/01441647.2021.1903841

Elmqvist, T., et al. (2018). *Urban planet: Knowledge towards sustainable cities*. Cambridge University Press.

Fergnani, A. (2019). Mapping futures studies scholarship from 1968 to present: A bibliometric review of thematic clusters, research trends, and research gaps. *Futures, 105*: 104–23

Fergnani, A. (2022). Corporate Foresight: A New Frontier for Strategy and Management. *Academy of Management Perspectives, 36*(2), 820–844. https://doi.org/10.5465/amp.2018.0178

Glaeser, E. L. (2011). *Triumph of the city: How our greatest invention makes us richer, smarter, greener, healthier, and happier*. Penguin Press.

Hammes, C. C. (2023). *Colonialism and the development of the modern South Asian City: Some theoretical considerations* (Vol. 1, pp. 1–19). Routledge eBooks. https://doi.org/10.4324/9781003355359-1
Iksan, M., Soong, J.-J., & Anandi, M. F. (2023). Jakarta in a time of political and economic change: Capital transformation, City Resilience, and Collaborative Governance. In: G. I. Neszmélyi & P. Koundela (Eds.), *Cities in Asia* (pp. 12–37). https://doi.org/10.29180/978-615-6342-64-5
King, A. D. (2004). *Spaces of global cultures: Architecture, urbanism, identity.* Routledge.
Kundu, D., & Pandey, A. K. (2020). World urbanisation: Trends and patterns. In: D. Kundu, R. Sietchiping & M. Kinyanjui (Eds.), *Developing national urban policies* (pp. 13–49). Springer. https://doi.org/10.1007/978-981-15-3738-7_2
Lees, L., Shin, H. B., & López-Morales, E. (2016). *Planetary gentrification.* Polity Press.
Liu, J. (2021). Asian city futures: Research to help inform spatial form and health. *Cities & Health, 5*(3), 215–223. https://doi.org/10.1080/23748834.2021.1970413
Mumford, L. (1961). *The city in history: Its origins, its transformations, and its prospects.* Harcourt, Brace & World.
Pacione, M. (2009). *Urban geography: A global perspective.* Routledge.
Park, R. E. (1915). The city: Suggestions for the investigation of human behavior in the city environment. *American Journal of Sociology, 20*(5), 577–612.
Pomeroy, G. M., & Mookherjee, D. (2023). Asian urban transformation: Emerging impressions and future research. In: D. Mookherjee, G. M. Pomeroy & L. T. T. Huong (Eds.), *Urban Transformational Landscapes in the City-Hinterlands of Asia. Advances in 21st Century Human Settlements* (pp. 497–512). Springer. https://doi.org/10.1007/978-981-19-8726-7_35
Rastouskaya, O. M. (2019). *Phenomenon of urban memory: Social-philosophical analysis.* Paper presented at the National Academy of Sciences of Belarus, Humanitarian Series. https://doi.org/10.29235/2524-2369-2019-64-1-33-40
Roy, A. (2009). Urban informality: Toward an epistemology of planning. *Journal of the American Planning Association, 71*(2), 147–158.
Sampson, R. J. (2012). *Great American City: Chicago and the enduring neighborhood effect.* University of Chicago Press.
Sassen, S. (2005). The global city: Introducing a concept. *Brown Journal of World Affairs, 11*(2), 27–43.
Scott, A. J., & Storper, M. (2015). The nature of cities: The scope and limits of urban theory. *International Journal of Urban and Regional Research, 39*(1), 1–15.
Singapore Ministry of Sustainability and the Environment. (2021). *Singapore Green Plan 2030.*
Syaban, A. S. N., & Appiah-Opoku, S. (2023). Building Indonesia's new capital city: an in-depth analysis of prospects and challenges from current capital city of Jakarta to Kalimantan. *Urban, Planning and Transport Research, 11*(1). https://doi.org/10.1080/21650020.2023.2276415
Townsend, A. M. (2013). *Smart cities: Big data, civic hackers, and the quest for a new utopia.* W.W. Norton & Company.
UN-Habitat. (2020). *The State of Asian Cities 2020.* United Nations Human Settlements Programme.
UN-Habitat. (2021). Overview: Cities and climate change initiative. Retrieved November 7, 2024 from https://unhabitat.org/programme/cities-and-climate-change-initiative
UNESCAP. (2019). Urbanization and sustainability in Asia and the pacific. United Nations Economic and Social Commission for Asia and the Pacific. Retrieved on November 7, 2024 from https://unhabitat.org/the-future-of-asian-pacific-cities-report-2019
United Nations. (2015). *Transforming our world: The 2030 agenda for sustainable development.* Retrieved on November 7, 2015 from https://sustainabledevelopment.un.org/post2015/transformingourworld/publication
United Nations. (2018). *World urbanization prospects: The 2018 revision.* Department of Economic and Social Affairs. Retrieved on November 7, 2024 from chrome-extension://efaidnbmnnnibpcajpcglclefindmkaj/https://population.un.org/wup/assets/WUP2018-Report.pdf

Van Noorloos, F., & Kloosterboer, M. (2018). Africa's new cities: The contested future of urbanisation. *Urban Studies, 55*(6), 1223–1241. https://doi.org/10.1177/0042098017707960
Watson, V. (2014). Co-production and collaboration in planning: The way forward? *Planning Theory & Practice, 15*(1), 62–76.
Zukin, S. (1995). *The cultures of cities*. Blackwell Publishing.

Rosilawati Zainol is a professor in urban analytics at the Department of Urban and Regional Planning, Faculty of Built Environment, Universiti Malaya. She currently holds the position of Deputy Dean for Undergraduate Studies for the faculty. She specialises in urban analytics, spatial analysis, sustainable tourism planning, and walkability.

Her research interests focus on the use of information technology in urban and regional planning, particularly GIS and spatial statistics. She has published numerous articles, book chapters, and books on sustainable urban and tourism development. Some of her recent academic works include studies on sustainable tourism governance, brownfield sustainability, child-friendly city assessment tools, and sustainable transportation systems.

Norzailawati Mohd Noor is a professor and Deputy Dean at the Kulliyah of Architecture and Environmental Design, International Islamic University Malaysia, whose expertise lies in urban planning technologies, particularly Geographic Information Systems (GIS) and remote sensing. With a Ph.D. and M.Sc from UTM, her research focuses on integrating advanced geospatial tools into urban planning to address challenges like urban sprawl, land use change, and disaster resilience. As Vice President of IGRSM and a registered Town Planner, she promotes the use of spatial technologies in policy-making and planning practices. Her work bridges academic research, technological innovation, and practical implementation in sustainable urban development.

Part I
Future Direction of Urban Planning: Complexity and Governance

Chapter 2
A Planning History Perspective to Understanding the Future of the Non-planning of the Urban Fabric in Manila, Philippines

Clusters of Future Studies: Past & Futures; Environmental Futures

Ian Morley

Abstract Despite the application of colonial and postcolonial urban planning models during the last 100 or so years, why has non-planning of urban land transpired in the Philippine capital city of Manila? In what ways has it existed as a direct response to the dynamics occurring within the broad societal setting, and/or flaws within planning systems? Furthermore, how has non-planning underpinned the enlargement of the country's administrative, cultural, and economic nucleus during past decades? How will it do so in the future? Thus, how can understanding of the urban future be shaped via expanding knowledge of past planning practices in the city? All in all, how can a discipline such as Planning History augment contemporary awareness of what Manila's, and indeed what Asia's urban future, could be?

Keywords Manila · Non-planning · Informal communities · Urbanisation · City management · Public policy

2.1 Introduction

In *The Oxford Handbook of Urban Planning* Weber and Crane (2015, 4) state that planning studies traditionally focus upon four fundamentals: the form of built and natural environments; the resolving of complex problems associated with urban growth and decline; the implementation and practice of planning systems; and, the creation of plans, processes, and institutions "that can induce, respond to, and manage change." Likewise, Ward et al. (2011, 231) observe that planning education now

I. Morley (✉)
Department of History, Chinese University of Hong Kong, Hong Kong, China
e-mail: ianmorley@cuhk.edu.hk

R. Zainol et al. (eds.), *Future of Cities in Asia*,
https://doi.org/10.1007/978-3-032-01016-2_2

commonly involves a historical component. In fact, they remark that there is now a compelling argument for making historical analysis be a core methodology of planning studies. Numerous reasons may be given as to why. These include, first, historical surveys assisting in conceptually defining the term 'urban', and also aiding the manufacture of scholarly lineage based on the form of the definition (Kwak, 2018, 53); second, it can inform of past planning models, namely their strengths and flaws; third, in conjunction with the prior point, it informs of planning errors associated with its conceptualisation and practice in the past so that they are not repeated in the present-day or future contexts; and, fourth, it abets new wisdom of how planning impacts different groups within urban society. In the light of the contemporary broadening of intelligence re city planning affairs in past, present, and future milieus, this work references the issue of non-planning in Manila, the Philippines. How has it come into being? How does non-planning's existence in the city align with the broader societal setting? More so, how will its presence affect Manila's future?

2.2 Methodology

The renowned planning historian Anthony Sutcliffe (1981, 65) wrote in *Built Environment* that urban planning by its very nature looks to the future. Yet, he adds, "planners need the past more than anyone, for without it they cannot draw on experience." In light of this remark, and by utilising core methods from the discipline of Planning History, this paper discloses how knowledge of the planned past can help inform of the urban present and future. Albeit with reference to Manila, the inquiry underscores how Planning History critically enlightens of the evolution of urban planning; it reveals a number of additional matters as well, namely ones vital to thoroughly comprehending the Philippine capital's present-day and future environmental form.

Broadly speaking, studies of town and city planning in history help to notify of administrative interventions that hitherto sought to regulate urban environmental challenges. Correspondingly, Planning History's nature highlights the strategies, agencies, and laws formed to improve the well-being of citizens. In alliance, planning historians have raised questions as to the role planners have had and how, even with the best of intentions, their plans sometimes exacerbated ongoing environmental, social, and economic predicaments in particular cities. Moreover, by engaging with specific places, and issues of spatiality and temporality, planning historians have exposed how planning practitioners have responded to problems relating to hygiene, housing, infrastructure, etc. (Hein, 2018, 7). Accordingly, Planning History has explained where, when, and why, planning systems have worked less efficiency than was originally intended.

Notwithstanding Sutcliffe's aforementioned judgement of the need for planners to have past-present-future temporal acquaintance, from the late-1960s efforts were made in the West to discredit centralised planning systems and their role in the shaping of built fabrics. As part of the advocation of what became known as the 'non-planning', the bypassing of existing planning regulations was encouraged so to instigate the betterment of urban dwellers' lives. This circumventing of bureaucracy was perceived to permit a more 'spontaneous' form of urban designing to occur. With the 'non-plan' being equated with freedom from official procedure, acts of 'non-planning' were judged to grant liberation from technocracy and its alleged negative imprint upon the lives of citizens. Yet, in the frame of this paper, a somewhat different definition is tendered. In this work the phrase 'non-planning' is not used to describe and/or symbolise planning's emancipation from state-composed regulation but, in contrast, it is employed as a term to describe the breakdown of an Asian state's planning framework. In other words, 'non-planning' is to be grasped within this inquiry as a failure of the Philippine planning system's capacity to safeguard convenience, beauty, public health, economic development, etc., for all urban citizens.

To explain non-planning in the Philippines, this paper puts attention upon the largest metropolitan place in the country, Manila.In 2020, the Philippine Statistics Authority recorded Metropolitan Manila as having a population of more than 13.48 million persons. This demographic total exposed that from the census of 2000 to the demographic survey of 2020, the metropolis's population grew by more than 3.5 million people. But, as dramatic and impactful as Manila's population growth has been during the early-2000s, it must be realised that its population has been swiftly expanding for more than 120 years. As Table 2.1 (see the next page) shows, since the first modern-era population enquiry, undertaken in 1903, Manila's population has grown exponentially; such growth is largely an upshot of in-migration. Evidently, with rapid growth throughout the 1900s, by about the year 2000 'Metro Manila', as it is locally known, had become a globally-renowned megacity. Unfortunately, such status was partly shaped by the countless slumlands that had formed within the city's bounds (Morley, 2018a, 17), *and* the high rate of poverty experienced by the local population. Even today, as the 2022 survey by the Social Wealth Station records, a significant proportion of the local populous 32%—self-rate their life[1] as occurring in alliance with poverty (Macasero, 2023). Shatkin (2004, 2470) with reference to the city's economic-environmental evolution labels the numerous poverty-ridden informal communities as 'forgotten places', and he sees the low-income residents of Metro Manila as being victims of a shelter crisis. In his view, changing patterns of employment in recent decades alongside rising property values "have translated into increases in the cost if (legal) housing, increased distances between places of employment and residence, and a deterioration of the urban environment. These changes have disproportionately affected the poor."

From the early-1900s, due to rapid urban growth, Manila was able to consolidate its standing as the Philippines' primate city (Morley, 2018b, 41–2). Nowadays, for example, its population scale is many times greater than the country's second most populated metropolitan place, Davao—population 3.33 million persons in 2020. In addition, given the history of rapid urbanisation within Metro Manila, today there

[1] The Philippine Statistics Authority in December 2023 set a poverty threshold for a family of five at 13,797 Pesos. See https://psa.gov.ph/statistics/poverty (accessed January 10, 2024).

Table 2.1 Population of the Manila metropolitan area recorded in national censuses since 1903

Year of national censuses	Population size (provide the unit here)
1903	219,928
1918	285,306
1939	623,492
1948	983,905
1960*	2,462,488
1970	3,966,695
1980	5,925,884
1990	7,947,392
2000	9,932,560
2010	11,855,975
2020	13,484,462

Source Philippine Statistics Authority (2020)
NB. Please note that from 1960, the national census recorded the population of the Manila metropolitan area, not just the City of Manila (as had formerly occurred, e.g. in 1903, 1918, 1939 and 1948)

are districts with seemingly preposterous population densities. Whilst nationally, population density is about 363 persons per square kilometre, in Metro Manila the population density is 21,765 persons. That said, as just indicated, the general population density of the metropolis pales in comparison to some of its highest concentrated localities, as Table 2.2 (see the next page) presents. Under such population pressure, and the paucity of available green space in the city, the degradation of the natural environment, the lack of social equity, and challenges to economic performance, the 2024 *Arcadis Sustainable Cities Index Report* has ranked Manila 93 out of 100 cities for the promotion and attainment of the United Nations' Sustainable Development Goals (SDGs). Given any familiarity with Planning History's historiography, e.g. of metropolises in both the West and East, the Arcadis evaluation comes as no surprise: empirical studies illustrate that any settlement's long-time failure to robustly manage urban growth, and to not overcome issues associated with rising population density and poverty, will inevitably trigger a host of both environmental and social actualities which now are considered an affront to the endorsement of SDGs; plus, conceptual studies of planning systems explain why planning in the context of rapid urban growth, regardless of place and time, can fail. The implications of this re the regulation of Manila's built fabric, it seems, has long been ignored by many public administrators just as it has in many other Asian megacities. That's why, claim Brennan and Richardson (1989), cities such as Delhi, Mumbai, Kolkata, Chennai in India, Dhaka (Bangladesh), Bangkok (Thailand), Jakarta (Indonesia), and Manila have inability to solve problems associated with urban services.

Given the previously discussed occurrence of rapid population growth, and, relatedly, the instigation of now deep-rooted social and environmental problems, evidence of state-led urban planning is ostensibly absent within numerous quarters of Metro

Table 2.2 Population density in 2020 of selected districts (known as 'barangays') in Metro Manila

Name of place	Population	Area (Km^2)	Population density (Km^2)
Barangay 67	4,083	0.0200	203,438
Barangay 85	4,032	0.0253	158,803
Barangay 95	5,543	0.0333	166,307
Barangay 99	6,310	0.1293	488,012
Barangay 101	15,358	0.1038	147,886
Barangay 111	2,661	0.1452	183,264
Barangay 152	6,218	0.0251	247,040
Barangay 221	3,353	0.0250	133,799
Barangay 230	2,364	0.0109	216,682
Barangay 310	13,602	0.0864	157,267
Barangay 375	3,909	0.0173	225,043
Barangay 745	2,015	0.0136	147,295

Source Philippine Statistics Authority (2020)

Manila. In many parts of the metropolis it is persons, usually of low social rank, who have organised the layout of buildings and open areas. Without formal planning training, they have organised open spaces and buildings by means of accretion; in other words, they arrange new structures and open areas around what already exists. Accordingly, in these locales, the risk of fire is high; when conflagration occurs, it is often devastating. Consequently, the history of fire and high-density living in Manila is deeply connected. Furthermore, the modern-era growth of Manila must be understood as a narrative of both urbanisation and densification, *and* the problems their alliance provokes. As a case in point, by as early as the mid-1900s, a lack of urban space was characteristic of many parts of the city. As a result, following subsequent urban growth and the further densifying of the built environment, by the mid-1950s in all but three districts of Manila, the amount of open space per capita was less than one square metre (Varias, 1955, 4) (see Table 2.3 on the next page).

Owing to the built environment evolving from the early-1900s with a lack of open space, Manila's population concentration had by the late-1940s already exceeded that of Manhattan, New York: 26,563 persons per square kilometres were recorded in Manila by the time of the 1948 census; circa that date, 23,887 persons were listed as residing in each square kilometre of Manhattan. As to why Manila had evolved with such a great population compactness it must be underscored that this was not because of an absence of a planning system. Laws existed to regulate urban development: a city plan was established in 1905; and, zoning plans had been composed by the 1930s. Significantly though, rules to manage urban development were often disregarded, *and* the practice of planning pre-national independence (in 1946) did not engage with the laying out of residential areas (Morley, 2023, 100–148). Hence, in districts such as Tondo, incredibly high urban densities were allowed to form before World War Two occurred and, post-war/post-independence, their population concentrations

Table 2.3 The area of public space (in square metres) within each district of Manila in 1955

District	Estimated population size	Estimated total area (in square metres) of public space	Average area of public space per resident (in square metres)
Binondo	22,565	9,600	0.43
Quiapo	29,768	4501.1	0.15
San Miguel	21,821	1.087.8	0.049
Sampaloc	280,624	2,387	0.008
Sta. Cruz	157,000	17,700	0.11
San Nicolas	43,113	980	0.022
Tondo	330,454	8.050	0.024
Paco	49,354	14,604	0.29
Pandacan	27,885	3,400	0.12
Sta. Ana	72,893	4,062.5	0.055
Intramuros	987	583,893.80	591.60
Ermita	14,922	798,551.50	53.50
Malate	71,175	564,495.78	7.90
Estimated open space per person in Manila (m^2)		1.79	

Source Varias (1955)

were allowed to further develop. Obviously, with the rise of high density living came a new scale of social-environmental problems. These concerns were discussed in the printed media by journalists. For instance, in 1938, *The Philippine Free Press* stated that a 'New Manila' must be formed, such were the daily challenges already faced by citizens. One writer, D.L. Francisco, recurrently harangued government officials for sitting back, doing nothing, and letting the urban density grow. Francisco (1941, 16) remarked that throughout Manila slumlands (see Fig. 2.1 on the next page) that are "dirty, insanitary places which stink to high heaven" were allowed to pot-mark the cityscape. In his view: "Among the most insanitary districts in Manila is Tondo". However, in "Sampaloc, Paco, Malate, Santa Ana, Singalong and Santa Mesa, similar filthy, insanitary and unhealthy conditions exist. Has anything been done?".

2.3 Key Concepts and Issues

In this part of the work brief attention is put upon the development of non-planned urban communities in Manila. The importance of contexts, such as densification, are discussed so to deepen grasp of dynamics that have shaped, and are still shaping, the built environment.

Fig. 2.1 An example of an informal community in the district of Tondo, Manila. *Source* The photograph was taken by the author during a fieldtrip in December 2022. Informal communities such as the one shown in the photograph were, from the early-1900s, regularly written about in the local printed press, and were the subject of local government attention

Knowing a city, especially its physical character, the disposition of local culture and peoples' way of life, as well as the make-up of local governance, are central to comprehending what a city planning system is and how it operates: studies in Planning History underscore the significance of this basic fact. Yet, matters of politics and culture are far from straightforward to grasp, and can involve transversal rationalities: hence, planning historians seek to know how and why planning 'impacts' peoples' way of life. In the setting of the Philippines, the complexity of understanding urban planning has been exacerbated by the sheer rate of urban development in recent decades. By way of illustration, and with reference again to Table 2.1 (see Page 24), Manila's built environment in 2020 housed almost fifteen times more people that what it did in 1948. Of note too, this expansion of the city's population and their growing compactness, akin to other cities in Asia has drastically affected levels of liveability. Shows Richards et al. (2017, 556), cities in Southeast Asia have developed with higher population densities, lower green space coverage, and less green space per capita than cities in the West. Manila, one of the world's densest cities, explicitly present compact living, and a paucity of open spaces. Given the enormous urban sprawl, and sheer scale of people within, in 2023 the Economist Intelligence Unit ranked the city 136 out of 173 for liveability (Simeon, 2024): on average, each worker in Manila loses approximately 117 hrs of time per year to traffic congestion. The 2023 Traffic Index found, of 387 cities surveyed in Asia and elsewhere, that

Manila is the slowest for traffic movement (Mateo, 2024). On top of these predicaments, the metropolitan area is prone to earthquakes and typhoons. It is estimated that about 170,000 residents of the City of Manila alone live in areas with high flood susceptibility, i.e. >1 m-high flood levels—an issue similarly experienced today in other large-sized Asian cities, e.g. Jakarta, Dhaka, Ho Chi Minh City (Vietnam), and Yangon (Myanmar). However, such environmental troubles are not exclusive to the inner-city quarters of Manila. Suburban-sited residents equally live their daily lives at risk of powerful natural phenomenon (see Fig. 2.2 on the next page). Therefore, what evidence now exists that state-led urban planning, other than peri-urban mega-projects, has brought improvement to the Filipino way of life?

In generic terms, numerous factors affect how an urban environment develops. One factor is city planning concepts and practices; they influence how an urban place can look and, in turn, be spatially arranged. But, in Metro Manila, non-planning has also had a profound effect on the character of the cityscape. Any visitor to the metropolis cannot ignore that throughout the built sprawl are unplanned, dense, and often decrepit areas. They have come into being due to the activities of landowners ignoring building regulations, the local and national government not stringently imposing regulations re issues of density and zoning, and squatters building upon available land (Abrams, 1966, 15). In view of this, Manila is now characterised by a dearth of open spaces. Only 0.2% of Metro Manila's land mass is green open space and, as a result, its built sprawl falls well short of the United Nations' recommendation of nine square meters of open space per resident (Zabala, 2023). That said too, as Table 2.3 (see Page 26) shows, many decades ago Manila had a lack of open space within its bounds. As such, the city's densification has to be acknowledged as not being a contemporary phenomenon. Then again, as the city grew throughout the 1900s and early-2000s not only did the urban sprawl expand and population size enlarge, already high-density districts were allowed to grow seemingly without check. Such a reality is evidence, notes Shatkin (2004), of the planning system 'forgetting' informal communities.

Studies of the character of Metro Manila's built fabric are not new, and have been undertaken for decades by intellectuals working within a range of academic fields. Local governments and national governmental offices too have for decades sought to understand the formation and development of urban districts, and the quality of life for people residing within them. In the late-1960s, for example, under the commission of Mayor Antonio Villegas, the book *Manila its Needs and Resources* (1967) was published. Amongst its major findings were that about 75% of households per annum faced financial challenges. In some parts of Manila, 'major financial difficulties' were annually experienced by about 50% of the population (Pablo, 1967, 229). Furthermore, in 1969, Aprodicio Laquian composed *Slums are for People.* In the book he discussed, among other things, the prevalence of people squatting on urban land. Other Asian cities are that time experienced this issue, e.g. Bangkok, Jakarta, Kuala Lumpur (Malaysia), Singapore, and Hong Kong (Aldrich, 2016).

The establishment of squatter communities, where persons build in haphazard fashion without the involvement of architects, planners, and civil engineers, were, said Laquian, the direct result of a low income economy. In other words, people

Fig. 2.2 A densely-packed suburban housing area in Metro Manila, photographed in early-2024, that is susceptible to flooding. *Source* The aerial photograph was taken by the author in February 2024

squatted because they did not have the financial means to compete in the housing market. Therefore, the only way for a low-income person to 'own' property was to find available land, and then to squat on it. Given the vast numbers of people presently undertaking squatting, it is estimated Manila now contains about 2,500 areas of informally developed land within its borders (see Fig. 2.3). However, as Laquian (1969, 76) shows, when persons incomes increase any advantages to squatting cease to exist. Of note too, what Laquian identifies in Manila is a culture of indifference between those who have, and do not have, high incomes. Simply put, those with money are willing, and happy, to reside in district built in accord with planning regulations so that cultural norms such as privacy can be upheld. Yet, those without high income or savings, are equally happy and willing to reside in non-planned areas even though, to those of higher social status, these communities are "overcrowded, smelly, insanitary and full of trouble." (Laquian, 1969, 177) Even if asked to move by public authorities off their squatting sites, Laquian finds that almost 75% of the slum dwelling populations would say "no".

Although Laquian (1969, 178) wrote about life in informal communities many decades ago, the issues/realities he mentions as to why non-planning occurs in Manila still exist; *his narrative of Manila's evolutionary planning story still has pertinence.* Thus, within informal communities less-affluent people are able to find economic, social, and psychological security even though life transpires "at bargain basement

Fig. 2.3 An informal settlement in the district of Sampaloc, Manila. *Source* The photograph was taken by the author on a visit to Manila in April 2024

rates." Those residing in slums typically do so to save money which will then be used for business activities, home improvement, and/or to fund their children's education; the people residing in the informal settlements want "to improve their houses, buy more things for the home, and generally, make life in the slums more tolerable. They usually don't want to move out of the slum area but are instead thinking of improving life there." (Laquian, 1969, 195).

2.4 Discussion of Implications

The widespread presence of informal communities in Manila, state Singh and Gadgil (2017, 3), has profound repercussions for those residing in them, i.e. it puts them at greater risk of insecure employment, vulnerability to economic exploitation, etc. Critically too, and now thinking with regard to environmental matters, more than 90% of Manila's informal communities are labelled as 'high dense areas'. More exactly, asserts Berner (1997), much high-density informality has occurred in proximity to the city's commercial areas. In his opinion, informal urban development transpires as a direct consequence of lowest income people realising that to live at distance from workplaces leads to daily wastes of time, energy, and, so, ability to earn money. Such a reality and its management, evidently, will affect the planning of Manila in the future. To be succinct, until the culture of low salaries, high land prices, and slow traffic movement is broken, and the state has full capacity to implement regulation to stop squatting, then non-planning will continue to thrivc. Moreover, the weaknesses of the state's place-based poverty alleviation strategies has, since the 1980s, made a challenging urban management situation become worse. Such approach to managing urban society is though not unique to Manila and has been evident in other Asian megacities (Kidokoro et al., 2022).

As mentioned previously, in the frame of this paper the term 'non planning' refers to an absence of planning regulation so that, ultimately, informal areas are widely produced within an urban environment (Palatino, 2012, 1). This, to reiterate with regard to Manila, is not because a planning system is missing. Three reasons underscore the existence of a city planning system in Manila. First, planned districts were laid out during the city's colonial and postcolonial past (Morley, 2023). From 1928, with Act No. 3482, the local administration as part of its governmental remit was able to undertake urban planning (Morley, 2019, 33). Second, the City of Manila Government presently, as it has done in the past, employs "Planning Professionals and Personnel dedicated to work in collaboration with partner stakeholders in formulating integrated multi-sectoral policies, plans and programs". Albeit now, these persons are employed to accomplish "Magnificent Manila: Green, Resilient, Inclusive, Smart and Sustainable." Third, the City of Manila Government presently has a City Planning and Government Office. It, as articulated via its Citizens Charter (2019, 87), has a mandate to "(1) formulate economic, social, physical and other development plans and policies for consideration of the local government development council; (2) conduct continuing studies, researchers, and training programs necessary to evolve

Fig. 2.4 An aerial photo, from February 2024, of Metro Manila's built sprawl. *Source* The photograph was taken by the author on a fieldtrip to Manila. It shows that within the metropolises' bounds different urban morphologies and verticality comprise the contemporary built fabric. The physical character of the metropolis is not only a reflection of socio-economic realities on the ground, but the failure of public authorities to formally plan so that urban life can benefit everyone irrespective of socio-economic status

plans and programs for implementation; (3) integrate and coordinate all sectoral plans and studies undertaken by different functional groups of agencies; (4) monitor and evaluate the implementation of different development programs, projects and activities in the local government units concerned and accordance with the approved development plan; (5) prepare comprehensive plans and other development planning documents for consideration of the local development council; (6) analyze the income and expenditure patterns and formulate and recommend fiscal plans and policies for consideration of the finance committee of the Local government unit concerned; (7) promote people participation in development planning within the local government unit; (8) exercise supervision and control over the secretariat of the local development council; and (9) exercise such powers and perform such others functions and duties as maybe prescribed by the law or ordinance, among which is the administration and enforcement of the zoning regulation in the City of Manila." The City Planning and Government Office will (2019, 87), the aforesaid Charter adds, "help fulfil the collective aspirations for the city and contribute in sustaining the goals and objectives of elevating the quality of life of its citizenry." But, to cite Murphy and Hogan (2012, 10), Manila (see Fig. 2.4) has evolved to be "one of the world's most fragmented and un-public of cities." It is also, as the 2024 *Arcadis Sustainable Cities Index Report* confirms, now one of the world's least effective cities concerning the application of SDGs.

To appreciate how the built fabric in Manila during recent decades has been shaped, attention must be placed upon three agencies: the national government; the local government; and international aid-lending organisations, e.g. the World Bank. Given the formation and enlargement of the Philippine economy, a significant part of which focuses upon Metro Manila, pressure for land development as part of domestic economic expansion has been forceful. With persons migrating en masse into the metropolis for the purpose of employment, those on the lower incomes have experienced insecurity of housing tenure which has exacerbated the long-standing existence of informal settlements. As Shatkin (2004, 2474) identifies, the existence of informal communities relies "on government tolerance of their presence on public land, or on private-sector actors who are waiting for future land appreciation, and therefore decide to forestall the cost and legal complication of eviction, or perhaps gain revenue in the short term by illegally renting land to the poor for housing." These circumstances, combined with what Shatkin (2000) notes as the selective memory of contemporary urban planners, has led to a land development approach in which the housing market is perceived to find its own solutions to the problems the planning system has formerly generated (see Fig. 2.5 on the next page). Of significance too, at least for those in the city's public offices, the generic interventionist policy approach to land and housing management, on one hand, has enabled planning practices to be devised which cost the cash-strapped city government relatively little money, and, on the other hand, it means that in Manila where the land-owning elites have for decades wielded political influence (Mouton & Shatkin, 2020, 403–22), politicians are not able to hinder their personal and political interests nor conflict with their role to ensure the provision of new housing for the public. Then again, such an approach has permitted a societal development paradox to emerge. On one hand, the national and local government's strategy has led to the building of infrastructure, of developing more efficient land use (given SDGs), of encouraging private sector funding, etc., but, on the other hand, whilst public policy has offered lip service to grasping and solving the needs of the lower social classes, in truth it has operated with confrontation to many citizens' daily needs (Shatkin, 2004, 2477). In summarising the existence of the public policy versus public needs inconsistency, Shatkin highlights the 'loss of institutional memory' with regard to comprehending the conditions of life for the urban poor notwithstanding governments for decades placing poverty alleviation at the heart of their societal betterment strategy (2479). In fact, despite the near-constant reference to social betterment within public policy by local and national governments, critically, the percentage of Manila's population in informal housing since the 1960s has not lessened; in truth, it has risen. So, what efficacy has state-led planning had in reconciling the well-being of the lower social orders? Does non-planning's existence in Manila show it, to be blunt, is little?

Fig. 2.5 Gentrification in the inner district of Malate. *Source* The photograph was taken by the author. Such high-rise development has led to a sharp rise in land values and, in consequence, the economic exclusion of less well-off social groups. The Urban Displacement Project now lists the area seen in the photograph (taken January 2024) as one of 'exclusion'

2.5 Future Directions

If recognising the practice of urban planning as a basic activity of the modern state, then decisions made as to the nature of urban designing have serious outcomes for those residing in an Asian metropolis such as Manila. Indeed, if governance is formed for the purpose of improving urban liveability, then unless it engages with all citizens and community groups a 'messy urbanism' can transpire. In the words of Hou and Chalana (2016, 9), "the making of the informal city and urban messiness does offer lessons for urban planning, policy, and governance." They add, there is now a growing need to acknowledge informal development as another way of urban planning and that it is "inseparable from formal planning." Yielding outcomes for specific groups in society, a place such as Metro Manila exposes the need to conceptually rethink how informal communities are formed, and their relationship to both the formal and informal systems of arranging the built fabric. Emphasize Hou and Chalana, informal planning should no longer be judged as counterhegemonic/ negative. Rather, the existence of non-state led planning—non-planning—does more than cite subaltern groups in dealing with the flaws of governance. Therefore, taking forward Hou and Chalana's logic, is it now time to take up Planning History so to rethink about how planning systems have worked, and failed, in Manila? More specifically, should we now rethink how in the past and present-day contexts those who live in informal settlements utilise non-planning so as to be active and engaged members of urban society? In truth, is the historical production of 'non-planned urbanism' an upshot of their efforts to be engaged and active citizens? If so, isn't there now a great need to reassess urban development outside the formal, and arguably ineffective, planning system? Can rethinking the history of planning, indeed knowing more of the planning past, help guide the planning future?

All in all, the historiography of non-planning as a component of city planning narratives is much smaller than it should be. As one of the enablers of urban growth is the capacity of planning professionals to inform of the concepts and practices that arrange the urban environment and so, in theory, provide for good health, economic growth, adequate housing provision, convenience, beauty, etc., then, as this paper accentuates, the planning system in Manila has evolved with limited efficacy. Given this actuality, those living at the bottom of the urban economy have formed their own process of urban building given their needs for survival and self-advancement. Ultimately, against the backdrop of non-planning practice, the paper indirectly asks two key questions: first, how can planning approaches attain benefits for those most in need of state assistance? (Shatkin, 2004, 2482); and, second, can the Asian city planning future be better shaped by being more cognisance of the urban past and failures within planning systems?

2.6 Conclusion

The discipline of Planning History—not formally taught in the Philippines—can assist both scholars and planning practitioners can rethink what planning is, how it works, and who benefits from it. As Carola Hein, the current President of the International Planning History Society outlines in *The Routledge Handbook of Planning History* (2018), knowing the past of planning can perform numerous intellectual tasks. These include: helping better understand planning's historical influence on our cities and nations; and, expanding grasp of the downsides of planning practice, as well as the need for novel approaches in order to resolve ongoing socio-economic and environmental challenges. Without such wisdom, Hein asserts, the ability of governments to elevate citizens' resilience to environmental and economic predicaments will be undermined. In the Philippines, where in 2023 22.4% of the nation cannot afford basic food and non-food needs (Philippine Statistics Authority, 2023), the capacity of the state to work with non-planning activity will be critical to the well-being of millions of people, i.e. those in *and* beyond the boundaries of Metro Manila. As such, there is much to learn of the history of urban planning so to better guide the urban future.

In this paper, lessons from Manila's planning history can be learned for the benefit of other Asian cities. Metropolises including Jakarta, Bangkok, Dhaka, and Mumbai have all experienced rapid growth during the past decades, and they now contain significant numbers of people living in poverty and/or in informal communities. Thus, how have their planning systems sought to improve liveability? What, in planning terms, has worked and not worked? Do their planning systems, in the past and also today, share common elements with Manila? If so, in what ways and why? But, most importantly, if liveability within their bounds is not good for many then why has the planning system not worked as effectively as it should? Common experiences can provoke broad debate as to what the Asian present has become. It can inform too of, likely, where the Asian urban future will go.

Highlights

- The significance of non-planning in shaping a capital city's urban form.
- Planning History as an intellectual field with distinct methods and perspectives to better grasp dynamics influencing contemporary and future urban development in Asia.
- A large-sized, and dense, city presenting itself as largely unplanned; an upshot of weak governance, the construction of slumlands, etc.
- The rise of alternative planning, i.e. non-planning, as a consequence of the failure of the mainstream planning system.

References

Abrams, C. (1966). *Squatter settlements. The problem and the opportunity*. Department of Housing and Urban Development.

Aldrich, B. C. (2016). Winning their place in the City: Squatters in Southeast Asian Cities. *Habitat International, 53*, 495–501.

Berner, E. (1997). The Metropolitan Dilemma. Global Society, localities, and the struggle for Urban Land in Manila. In: O. Ayse & P. Weyland (Eds.), *Space, culture and power. New identities in globalizing cities* (1st ed., pp. 98–116). Ze Books.

Brennan, E. M., & Richardson, H. W. (1989). Asian megacity characteristics, problems, and policies. *International Regional Science Review, 12*(2), 117–129.

City Planning and Development Office. (2019). *Citizen's Charter*. City of Manila Government.

Francisco, D. L. (1941, November 15). There are no funds! *Philippine Free Press*, 16.

Hein, C. (2018). The what, why, and how of planning history. In C. Hein (Ed.), *The Routledge handbook of planning history* (pp. 1–10). Routledge.

Hou, J., & Chalana, M. (2016). Untangling the 'Messy' Asian City. In M. Chalana & J. Hou (Eds.), *Messy urbanism. understanding the "Other" cities of Asia* (1st ed., pp. 1–21). University of Hong Kong Press.

Kidokoro, T., Matsuyuki, M., & Shima, N. (2022). Neoliberalization of urban planning and spatial inequalities in Asian megacities. *Cities, 130*, 1–15.

Kwak, N. (2018). Understanding 'Urban' from the disciplinary perspective of history. In D. Iossifova, C. N. H. Goll & A. Gasparatos (Eds.), *Defining the urban: Interdisciplinary and professional perspectives* (1st ed., pp. 53–62). Routledge.
Laquian, A. (1969). *Slums are for people*. DM Press, Inc.
Macasero, R. (2023, February 7). In 2022, more people felt poor in regions outside Metro Manila—SWS. *Rappler*. https://www.rappler.com/nation/more-people-felt-poor-regions-than-metro-manila-sws-survey-january-december-2022/
Mateo, J. (2024, January 14). Traffic index: 25 mins, 30 s to Travel 10 kms in Metro Manila. *Philstar Global*. https://www.philstar.com/headlines/2024/01/14/2325664/traffic-index-25-mins-30-secs-travel-10-kms-metro-manila
Morley, I. (2018a). City profile: Manila. *Cities, 72*, 17–33.
Morley, I. (2018b). *Cities and Nationhood: American imperialism and urban design in the Philippines, 1898–1916*. University of Hawai'i Press.
Morley, I. (2019). *American colonisation and the city beautiful. Filipinos and planning in the Philippines, 1916–35*. Routledge.
Morley, I. (2023). *Remodelling to prepare for independence. The Philippine commonwealth decolonisation, cities and public works, c. 1935–46*. Routledge.
Mouton, M., & Shatkin, G. (2020). Strategizing the for-profit City: The state, developers, and urban production in Mega Manila. *Environment and Planning A: Economy and Space, 52*(2), 403–422.
Murphy, P., & Hogan, T. (2012). Discordant order. Manila's neo-patrimonial urbanism. *Thesis Eleven, 112*(1), 10–34.
Pablo, L. (1967). *Manila its needs and resources*. City of Manila Government.
Palatino, M. (2012). *Blogging the city and spaces of dissent in the Philippines*. Southern Voices Printing Press.
Philippine Free Press. (1938, April 30). Building a New Manila, 4.
Philippine Statistics Authority. (2020). Available at https://rssoncr.psa.gov.ph/content/highlights-2020-censuspopulation-and-housing-city-manila
Philippine Statistics Authority. (2023). Preliminary 2023 First Semester Official Poverty Statistics. https://psa.gov.ph/statistics/poverty
Richards, D. R., Passy, P., & Oh, R. R. Y. (2017). Impacts of population density and wealth on the quantity and structure of urban green space in tropical Southeast Asia. *Landscape and Urban Planning, 157*, 553–560.
Shatkin, G. (2000). Obstacles to empowerment: Local politics and civil society in Metropolitan Manila, the Philippines. *Urban Studies, 37*(12), 2357–2375.
Shatkin, G. (2004). Planning to forget: Informal settlements as 'Forgotten Places in Globalising Metro Manila.' *Urban Studies, 41*(12), 2469–2484.
Simeon, L. M. (2024, September 29). World's most liveable cities: Metro Manila Slips to 236th. *Philstar Global*. https://www.philstar.com/headlines/2023/06/28/2277077/worlds-most-liveable-cities-metro-manila-ranking-slips-136th
Singh, G., & Gadpil, G. (2017). *Navigating informality. Perils and prospects in Metro Manila's Slums*. The World Bank.
Sutcliffe, A. (1981). Why planning history? *Built Environment, 7*(2), 64–67.
Varias, A. (1955). Public Plazas and open spaces in the city of Manila. *The Real Estate Digest, 4*, 4–5.
Ward, S., Freestone, R., & Silver, C. (2011). The 'New' planning history: Reflections, issues and directions. *Town Planning Review, 82*(3), 231–262.
Weber, R., & Crane, R. (2015). Planning as scholarship: Origins and prospect. In R. Weber & R. Crane (Eds.), *The Oxford handbook of urban planning* (1st ed., pp. 3–20). Oxford University Press.
Zabala, H. (2023, February 11). Claiming back the city. *Manila Standard*. https://manilastandard.net/?p=314303715

Ian Morley is a Professor in the Department of History at the Chinese University of Hong Kong. Prof. Morley has published widely on the design of built environments in the Philippines. He is an editorial board member of *Planning Perspectives* and *World History Bulletin*, as well as a council member of the Epidemic Urbanism Initiative (EUI). He is also Vice President of the International Planning History Society. He regularly contributes op-eds in the Philippine printed media.

Chapter 3
Urban Amnesia: Spatial Traces of Prolonged Occupation in Gaza City, Palestine, and the Future of Other Conflict-Prone Cities in Asia

Clusters of Future Studies: Past & Futures; Humanity at the Limen

Fatih Eren

Abstract This conceptual chapter examines the concept of 'urban amnesia', which refers to the forgotten information within a city. The focus is on Gaza, which has faced continuous occupations and conflicts over the past 75 years, unlike many other cities that have enjoyed peace since World War II. The chapter aims to map the spatial markers of Gaza's long-term occupation through urban amnesia and draw conclusions about the future of Asian cities that might face similar threats, such as Kabul (Afghanistan); Lahore (Pakistan); Srinagar (India); Yangon (Myanmar); Bangkok (Thailand); Hanoi (Vietnam); Seoul (South Korea); Tehran (Iran); Beirut (Lebanon); Jakarta (Indonesia); and Bishkek (Kyrgyzstan). Data were collected through a literature review and document analysis, employing thematic content analysis of eight significant urban spaces in Gaza, using the codes 'unforgettable-event' and 'spatial trace of experience.' The chapter begins with a theoretical exposition of urban amnesia, followed by an exploration of Gaza's unique context. Insights are derived regarding the future of Asian cities in potential conflict scenarios, emphasising the need to safeguard urban elements that connect them to their land, culture, and identity. The end of Gaza's occupation represents a crucial hope for a peaceful future for conflict-prone cities in Asia, highlighting the moral and legal responsibilities of humanity.

Keywords Palestine · Gaza · Urban amnesia · Occupation · Asian Cities · Future

F. Eren (✉)
Istanbul Technical University, Faculty of Architecture Department of Urban and Regional Planning, Istanbul, Türkiye
e-mail: feren@itu.edu.tr

R. Zainol et al. (eds.), *Future of Cities in Asia*,
https://doi.org/10.1007/978-3-032-01016-2_3

3.1 Introduction

> The destruction of the past, or rather of the social mechanisms that link one's contemporary experience to that of earlier generations, is one of the most characteristic and mysterious phenomena of the late twentieth century. Many young men and women at the century's end grew up in a sort of permanent present lacking any organic relation to the public past of the times they live in.
>
> (Hobsbawm, 1994: 3)

This section discusses the urban amnesia experienced by the city of Gaza in Palestine as well as relates to the future of Asian cities, such as Kabul (Afghanistan); Lahore (Pakistan); Srinagar (India); Yangon (Myanmar); Bangkok (Thailand); Hanoi (Vietnam); Seoul (South Korea); Tehran (Iran); Beirut (Lebanon); Jakarta (Indonesia); and Bishkek (Kyrgyzstan). The events that occurred in Palestine, particularly in Gaza, after World War II are super-political actions that wound consciences on a humanitarian-moral level and have global impacts. When evaluated within a broader framework, it is evident that the destruction and tragedy Gaza has experienced over the last 75 years carry significant lessons for the peaceful future of Asian cities. This is because after the completion of the occupation in Gaza, there is a high probability that war and conflict will shift to the Asian continent, which has once again become the centre of gravity for the world (see Fig. 3.1).

The history of Palestine is marked by enduring struggles for survival, loss of land, and erosion of identity and culture. Although it has been a focal point for numerous civilisations since ancient times, the twentieth century has been particularly challenging. The historical Palestinian territories, situated between the Jordan River and the Mediterranean Sea and governed by the Ottoman Empire for around 400 years, began to be incrementally occupied following the empire's decline. From

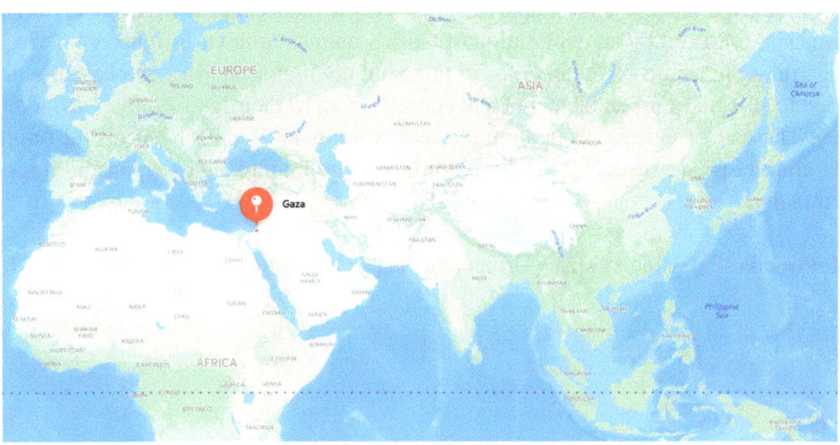

Fig. 3.1 The location of Gaza City, Palestine, within the Asian continent. *Source* Yandex Maps (2024)

1917 onwards, these territories experienced a significant, organised influx of Jewish immigrants. On 14 May 1948, the National Jewish Council proclaimed the creation of a new state on these historical lands. This led to the forced expulsion of thousands of Palestinians from their homes, an event known as the 'Nakba' (Great Catastrophe). The Six Day War in 1967 resulted in the occupation of Gaza and East Jerusalem, further disrupting Palestinian daily life and initiating a painful struggle for existence, characterised by land loss, cultural and identity erosion, and human rights violations. This book chapter narrates the story of Palestine through the lens of urban amnesia, focusing on the city of Gaza.

The study introduces the concept of urban amnesia and examines it specifically in the context of Gaza city. It addresses the losses in Gaza's urban fabric and collective memory resulting from wars and conflicts. The transformation and losses experienced by the city are detailed through eight important urban spaces that house Gaza's collective memory: Gaza Great Omar Mosque, Gaza Port, Gaza Traditional Marketplace, Gaza Palestine Square, The Islamic University of Gaza, Gaza Al-Shifa Hospital, Gaza Beach, and Gaza Hotel Museum. At the end of the study, inferences are made for Asian cities under potential war-conflict threats, and concrete urban proposals are presented.

3.2 Methodology

In the research, qualitative data was gathered through literature review and document analysis methods. To explore the spatial traces of lived experiences in Gaza City, Palestine, keywords such as 'Gaza', 'urban amnesia', 'collective memory', 'occupation', 'war', and 'cultural heritage' were used in integrated search engines and databases like Scopus and Web of Science. The reliability and validity of the chosen sources were ensured by their publication in academic journals, the peer-review process, and the authors' expertise in the field.

The data collected was analysed using the 'content analysis method'. This method facilitates a detailed evaluation of qualitative data (texts) and enables an objective and systematic examination of written materials (Çalık & Sözbilir, 2014; Dinçer, 2018; Falkingham & Reeves, 1998; Tavşancıl & Aslan, 2001). The content was categorised into two main themes (unforgettable-exemplary event, spatial trace of experience) and four sub-themes (physical destruction, symbol spaces-artifacts-buildings, tension-conflict-attack point, urban memory). It was analysed by interpreting the data and identifying spatial evidence of Gaza's urban amnesia. The content analysis was extended using the same keywords and thematic codes for the eight most notable and physically lost urban spaces in Gaza's urban fabric and collective memory.

3.3 Key Concepts and Issues

3.3.1 Urban Amnesia

Urban amnesia refers to the erosion or forgetting of a city's historical and cultural memory (Rastouskaya, 2019; Wicke, 2023). This phenomenon involves a collective memory loss driven by the rapid transformation of a city through various local and global influences (Crinson, 2005). Factors such as globalisation, rapid urbanisation, migration, and economic-political ambitions, along with extraordinary events such as war, conflict, violence, chaos, crises, and disasters, contribute to the disappearance or neglect of structures, spaces, and cultural elements that compose the city's fabric (Nagle, 2017). This leads to a disconnection from the city's past, loss of its unique character, disruption in the transfer of cultural knowledge across generations, and weakening of urban identity, social memory, and a sense of belonging (Chang, 2020; Farrar, 2011).

There are diverse opinions on urban amnesia in the literature. One view posits that urban change is inevitable due to modernisation and technological progress (Qiu Sun, 2021). Advocates of this perspective acknowledge some degree of urban amnesia but prioritise the city's development and adaptation to modern requirements (Barreiro León, 2017). This forward-looking approach is often criticised for neglecting the preservation of historical and cultural heritage (Rajagopalan, 2012). An opposing view emphasises that preserving urban memory is essential for maintaining the city's identity and continuity (Sargın, 2004; Wang and Wang, 2016). This method is largely endorsed by urban historians, city planners, and cultural heritage specialists, yet it faces criticism for possibly hindering urban growth due to an overemphasis on preservation (Larkham, 2002). Social scientists and activists often highlight that urban amnesia disproportionately impacts disadvantaged groups' memories and living spaces (Eren & Yıldız Keçer, 2023; Lewis-McCoy et al., 2023), though they face criticism for overlooking the potential benefits of urban development and change (Meerow et al., 2019). The techno-optimist perspective argues that digital technologies offer new opportunities for preserving urban memory (İnce, 2014; Gonzaga, 2019; Eren et al., 2023). Technology experts and some city planners support this approach, though it is criticised for claiming that virtual preservation cannot fully replace physical preservation (Silberman & Purser, 2012). Meanwhile, the palimpsest theory views cities as layered texts, where each layer retains remnants of the previous ones (Diener & Hagen, 2018). This perspective, held by some urban theorists and architects, is criticised for being too abstract and impractical (Engbersen, 2001; Wyly, 2024). The cultural ecology-based perspective suggests evaluating urban amnesia within the broader ecological system of the city (Agaeva, 2021), although it faces criticism for ignoring economic and political realities (Loftus, 2012).

Scholarly reviews reveal that urban amnesia prominently arises in cities with particular characteristics:

- Historical and touristic cities with rich cultural heritage, such as Jakarta, Beijing, Bali, Rome, Athens, Cairo, Venice, and Prague, face risks of authenticity loss due to modernisation, tourism, or economic pressures,
- Industrial cities like Tokyo, Seoul, Shanghai, Manchester, and Detroit, which have shifted to the service sector, experience transformations in old industrial areas,
- Cities undergoing reconstruction due to war or natural disasters, such as Kabul, Kathmandu, Manila, Beirut, New Orleans, and Hiroshima, encounter significant physical and social changes,
- Global financial and technology hubs like Shanghai, Hong Kong, Singapore, London, New York, San Francisco, and Bangalore face rapid economic growth and changes in local identity under global influences,
- Former colonial cities such as Mumbai, Kolkata, Hanoi, and Algiers, which are transitioning from colonial pasts, grapple with cultural identity issues,
- Multicultural metropolises like Kuala Lumpur, Dubai, Istanbul, Toronto, Amsterdam, Bangkok, and Sydney, which receive intense immigration and experience cultural integration processes, also frequently discuss urban amnesia.

In cities affected by war or occupation, urban amnesia embodies multi-layered and complex implications (Hasan, 2019). Physical destruction leads to significant losses of the original urban fabric (Bevan, 2007), and the collective memory of war or occupation risks being erased during reconstruction (Graham, 2008; Susser & Schneider, 2003). Cities like Kabul, Phnom Penh, and Dresden serve as examples of such risks. Occupied cities may experience changes or suppression of cultural identity due to new regimes reshaping them (Blockmans, 2003; Grodach, 2002), as seen in Seoul under Japanese occupation, Urumqi under Chinese control, and Lhasa post-Tibet occupation. Determining which memories to preserve or forget becomes a crucial and politically charged topic in cities recovering from war or occupation (Bleibleh & Awad, 2020; McDowell, 2016). Debates, such as those in Japan over the Hiroshima bombings or in Germany regarding Nazi-era structures, exemplify these challenges. Facing past traumas head-on to achieve social reconciliation involves navigating dilemmas on how to balance remembering or forgetting the traumatic past (Adams, 2011; Goebel & Keene, 2011; Verpoest et al., 2020). The dilemmas experienced in post-war Sarajevo and post-1975 Dili illustrate the complexities of handling urban memory.

Wars and occupations significantly lead to population movements and forced migrations in cities (Ahonen et al., 2020; Huff, 2014; Hugo, 2006), resulting in a complex process of adaptation for new groups and urban amnesia (Klein, 2008; Sugrue, 2017). Symbolic importance of specific structures or spaces can increase or decrease (Burch & Smith, 2007; Silver, 2010), exemplified by the Auschwitz Monument in Hong Kong or Democracy Square in Gwangju. Hence, in war-affected cities, urban amnesia is a multifaceted issue involving physical, social, cultural, and psychological dimensions, with sensitive balances between memory and forgetting closely tied to social reconciliation and future outlooks.

3.3.2 Urban Amnesia and Spatial Traces of Prolonged Occupation in Gaza City, Palestine

Since 1948, Gaza City, historically marked by wars, conflicts, and occupations, harbours significant urban spaces that are deeply embedded in the collective memory of its inhabitants. Over the past 75 years, these spaces have become witnesses to and have been altered by the city's political, economic, and sociological turmoil. The occupation movement that commenced on October 7, 2023, has notably resulted in the deaths of over 40,000 individuals, left many injured, and has triggered catastrophically severe issues such as famine, leading to widespread physical destruction across Gaza. Previously home to 600,000 residents, Gaza now faces extensive devastation, with approximately 74.3% of its buildings either damaged or demolished.

Eight key locations in Gaza City, which have intimately experienced the city's events and possess profound historical significance, are as follows: Great Omari Mosque, Gaza Port, Traditional Gaza Marketplace, Gaza Palestine Square, Islamic University of Gaza, Al-Shifa Hospital, Gaza Beach, and Mathaf al-Funduq (Gaza Hotel Museum). These locations embody Gaza's urban memory and are elaborated upon below.

3.3.2.1 Great Omari Mosque

This landmark has been embedded in the collective consciousness as a sacred place of reverence. Prior to Christ, it served as a temple for local religions, and by the 500s AD, it was utilised as a church by Christians (Shehada, 2020). From the 700s AD, it functioned as a significant mosque for Muslims for 14 centuries. Following the Six-Day War in 1967, the occupiers took over the mosque's administration (Feldman, 2007). During the First Intifada (1987–1993), it became a central venue for protests and clashes. After the Oslo Accords in 1994, Palestinians regained administrative control. Throughout the Second Intifada post-2000 and during the Cast Lead, Pillar of Defence, and Protective Edge Operations, the mosque suffered numerous bombings and demolitions but was consistently repaired by the citizens of Gaza. Emblematic of Gaza's resistance and resilience, the Great Omari Mosque was demolished once again by occupying warplanes in December 2023.

3.3.2.2 Gaza Port

With heightened strategic significance post-World War II, Gaza Port emerged as a primary entry point for numerous displaced individuals and migrants to Gaza (Hassouna & Nam, 2005; Smaling et al., 1996). The port sustained damage during the Suez Crisis in 1956. Following the Six-Day War in 1967, it fell under the control of occupying forces, resulting in prolonged restrictions on Palestinian access and fishing activities. Throughout the First Intifada (1987–1993) and the Second Intifada (2000–2005), the port area became a focal point for attacks, protests, and clashes. Post-2006, the occupiers imposed a stringent naval blockade on Gaza (Arraf & Shapiro, 2011). In 2010, the "Mavi Marmara" humanitarian aid flotilla, comprised of six ships with over 700 passengers from 36 countries, was attacked in international waters by the occupiers while attempting to deliver aid to Gaza (Topal, 2012). Consequently, due to the blockade, attacks, and various restrictions, Gaza Port's operational capacity has sharply declined, severely impacting the fishing industry, which is vital for the local economy. Gaza Port stands as a prominent symbol of Gaza's enduring struggle for freedom and economic autonomy.

3.3.2.3 Traditional Gaza Marketplace

Since 1948, the influx of people migrating to Gaza from the occupied Palestinian territories has significantly influenced the structure and economy of the historic Al-Zaviya and Saha markets, along with Fehmi Street. The 1970s and 1980s saw the market under occupation, which imposed severe restrictions on commercial activities and led to economic hardships (Abu-Amr, 2015). From the 1990s onward, heightened tensions, conflicts, restrictions, blockades, and economic sanctions have profoundly impacted the commercial landscape. Despite the challenges and adversities, efforts persist to sustain the commercial and social vibrancy of this traditional hub of trade and social life.

3.3.2.4 Gaza Palestine Square

Gaza Palestine Square, situated in the Rimal district (the heart of Gaza) has historically served as the primary assembly point for Palestinians who were compelled to relocate to Gaza from the occupied Palestinian territories since 1948. After each significant event in Gaza's history, residents have converged upon this square either to protest or to celebrate. Thus, it has evolved into a symbolic space where the collective emotions of the Gazan people, both jubilant and sorrowful, are most fervently expressed. Given its prominent role in the collective memory of Gazans, the square stands as a central node in Gaza's social and cultural landscape. Over the years, it has been a backdrop to both the greatest joys and deepest sorrows of the people. In December 2023, the square was bombed by the occupiers, who alleged that a network of tunnels lay beneath it.

3.3.2.5 The Islamic University of Gaza

Founded in 1978, the Islamic University of Gaza has become known for its active student engagement through protests and demonstrations. Following the Oslo Accords in 1994, both students and faculty experienced increased freedom to voice

Fig. 3.2 The Islamic University of Gaza, rendered unusable after airstrikes in October 2023. *Source* Palestinian Information Centre (2023)

their political views, sparking a resurgence of academic and cultural activities on campus (Al Shobakib et al., 2017). However, post-2005, the blockade on Gaza severely hindered the flow of materials and resources to the university, causing substantial financial difficulties. During subsequent military operations, the university's infrastructure, including laboratories and libraries, experienced considerable damage. International aid organisations have stepped in multiple times to repair the damaged buildings. The blockade and economic constraints have negatively impacted the university's academic partnerships with institutions worldwide (Milton et al., 2023). Tragically, in December 2023, the home of the university's Rector, Prof. Dr. Sufyan Tayeh, was directly targeted and bombed by occupying forces, leading to the deaths of the rector, his wife, and their children. In spite of these severe challenges, the Islamic University of Gaza has persevered in its educational mission and remains a pivotal academic institution in Gaza (see Fig. 3.2).

3.3.2.6 Gaza El Shifa Hospital

Established in 1946, El Shifa Hospital stands as Gaza's largest healthcare facility, originally designed to accommodate 700 beds. It has consistently been instrumental in providing treatment and care to the local population. With each offensive and incursion into the city, the hospital has faced an overwhelming influx of casualties, both injured and deceased. Over time, its emergency services and operating theatres have been forced to function nearly nonstop, straining its capacity as a medical centre. Chronic challenges have included constant shortages of electricity, medical

Fig. 3.3 Gaza Al-Shifa Hospital, once the largest healthcare complex, now reduced to a basic first aid centre after repeated attacks and sieges. *Source* Türkiye's National Public Broadcaster, TRT Haber (2024)

supplies, and personnel (Gilbert & Fosse, 2009). Power outages and the scarcity of fuel for generators have tragically led to the deaths of hundreds of premature infants who could not be sustained in incubators, as well as numerous kidney patients unable to receive dialysis. The World Health Organization has starkly described the hospital as having been transformed into a cemetery (Musa et al., 2023). Following bombings and substantial damage in November 2023, and subsequent sieges and raids by occupying forces in March 2024, numerous injured patients were compelled to evacuate and seek refuge elsewhere. Consequently, Gaza's primary medical complex has been downgraded to a rudimentary first aid station (see Fig. 3.3) (Alser et al., 2024).

3.3.2.7 Gaza Beach

Historically Gaza's primary spot for relaxation and social gatherings, the beach has typically served as a sanctuary for peace and calm within the city. Nevertheless, due to various attacks and invasion, it became a temporary haven for displaced residents who erected tents for shelter (Junka, 2006). Over time, this area tragically evolved into a site of civilian massacres observed by the international community. In June 2006, a naval shelling by the occupiers resulted in the deaths of 8 civilians and injuries to 32 others. Additionally, in July 2014, an airstrike by the occupiers claimed the lives of 4 Gazan children playing on the beach (see Fig. 3.4 on the next page) (Mohammed Asqool et al., 2023).

Fig. 3.4 Gaza Beach, where children were killed by airstrikes while playing, their names now written in the sand. *Source* From an article in the national daily news portal Ulusal (2022)

3.3.2.8 Mathaf Al-Funduq (Gaza Hotel Museum)

Situated on al-Rashid Street, this museum highlights Gaza's rich history and cultural legacy. It features artifacts that capture the historical and cultural significance of a city that has witnessed the rise and fall of over eight civilisations, including the ancient, Roman, Byzantine, and Islamic eras (Kosciejew, 2024). Inaugurated in 2008, the museum's collection was carefully curated over three decades, epitomising the dedication and cultural consciousness of a Palestinian woman (Almbaid, 2015). However, like 12 other museums in Gaza eventually subjected to bombing and destruction, the Gaza Hotel Museum was also targeted in November 2023 (As'ad, 2024). The assault on these cultural institutions, which serve as vital links for Gazans to their heritage, history, culture, and identity, underscores that Gaza is facing a systematic cultural war alongside the ongoing military conflict (see Fig. 3.5 on the next page).

The eight locations discussed above reveal much about the urban amnesia experienced in Gaza. This city finds itself ensnared in the relentless cycles of history. Every stone and street in Gaza carry the weight of millennia of memories, yet over the past 75 years, this ancient heritage teeters on the verge of evisceration. Time here acts not only as a continuum linking past and future but also delineates the fragile boundary between existence and nonexistence. Gaza's urban amnesia transcends mere physical destruction or street alterations; it signifies the fragmentation of an entire civilisation's essence and the dispersal of collective consciousness. Each explosion not only topples buildings but also obliterates the cultural fabric woven over centuries. Walking through the city's streets, one can sense different historical layers, yet these are suffused with pain and desolation. This forced oblivion in Gaza offers a reprieve from unbearable memories, yet the perils of forgetting loom large.

Fig. 3.5 The Gaza Hotel Museum, destroyed by bombing in November 2023 amid the ongoing cultural and military conflict. *Source* From an article by Saber (2024) in the international daily news portal Aljazeera

The erasure of a city's memory equates to the loss of its soul. The stones, walls, and corners of Gaza serve as a 'Time Regulation Institute' (Tanpınar, 1961). As Ahmet Hamdi Tanpınar observed, the disintegration of these institutes disrupts the perception of time, engulfing the city in a temporal limbo. To combat Gaza's urban amnesia, it may be crucial to preserve the city's essence, scent, sound, and texture through advanced technological means. This approach could forge a city memory impervious to both bombs and the passage of time. Ultimately, Gaza's urban amnesia represents not just the loss of a city's history but a segment of humanity's shared memory. Thus, remembering Gaza constitutes a moral obligation to humanity itself, transcending mere civic duty.

3.3.3 Urban Amnesia and Spatial Traces in Other Conflict-Prone Cities in Asia

3.3.3.1 Kabul (Afghanistan)

Kabul bears deep traces of urban amnesia and destruction after decades of war, occupation, and internal conflicts. The process that began with the Soviet invasion in 1979 continued through the civil war of the 1990s, followed by the Taliban era and the U.S. invasion, resulting in a constant transformation and devastation of the city (Cassar & Noshadi, 2015). In the old neighbourhoods, particularly around the Darulaman Palace, the scars of heavy bombardment and clashes are still observable; the bullet holes in the palace walls and the ruined corridors stand as indelible evidence of the past. In areas like Karte Seh and Karte Char in the west of the city, once vibrant apartment blocks have either been completely destroyed or left in a state of irreparable disrepair (Dogar et al., 2024). In the centre of Kabul, newly constructed shopping malls and glass-fronted buildings replace the old neighbourhoods and historic bazaars, obscuring the urban memory. This rapid and chaotic reconstruction has left indelible voids in both physical and social memory; the traces of the past intertwine with today's forgetfulness in the fabric of the city.

3.3.3.2 Lahore (Pakistan)

Lahore has experienced a multi-layered urban memory loss due to various invasions, conflicts, and social chaos throughout history. Especially during the partition of India and Pakistan in 1947, the city was shaken by massive migrations and ethnic violence (Cermeño & Mielke, 2023). In the narrow streets of the Old City, abandoned temples silently witness the communities that once lived here; doors covered with plaster and ruined courtyards reveal the erased layers of collective memory. The military coups and political instabilities of the 1970s and 1980s further fragmented Lahore's urban fabric, with inns and markets in the old neighbourhoods falling victim to unplanned urbanisation (Ristic & Frank, 2019). Today, the rise of newly developing areas like Gulberg and Defence overshadows the characteristic texture of old Lahore, disrupting the spatial continuity of the city. In this process, Lahore's historical identity and social memory oscillate between modernisation and forgetfulness.

3.3.3.3 Srinagar (India)

Srinagar carries the traces of urban amnesia and trauma under the shadow of decades-long military conflicts, occupation, and civil unrest. The traditional wooden houses and houseboats around Dal Lake have been significantly damaged due to frequent curfews, explosions, and fires (Winter & Panjabi, 2010). In the city centre, the buildings around Lal Chowk still bear the physical marks of bullet impacts and barricades.

The mosques and inns in the old city have been ruined due to repeated fires and security operations, with many traditional structures either completely destroyed or rendered non-functional. Military checkpoints and barbed wire, frequently encountered on the streets of Srinagar, have become a part of daily life, militarising the public spaces of the city (Panjabi & Winter, 2009). As the latest development, on May 6, 2025, India launched a military operation against targets in Pakistan and the Azad Kashmir region under its control. Following the attack, which resulted in numerous casualties, Srinagar Airport in the Jammu and Kashmir region of India was closed to civilian flights. In the tense atmosphere of Srinagar, which is under constant threat of war, social memory faces a continuous pressure and threat of oblivion, while the spatial fabric of the city bears the marks of trauma and memory loss.

3.3.3.4 Yangon (Myanmar)

Yangon deeply harbours the traces of military coups, ethnic conflicts, and political chaos within its urban landscape. While the city centre is filled with the grand buildings of the colonial era, these structures have been largely neglected since the military regime began in 1962 and during the public uprising in 1988, with some intentionally abandoned. Particularly around Sule Pagoda Road, old public buildings from the British period have fallen into ruin due to neglect and vandalism, while some have been repurposed for new commercial projects, losing their identity (Logan, 2016a, 2016b). The increased violence and protests following the 2021 coup have led to the re-establishment of military checkpoints and barricades, militarising public spaces. Temples in old neighbourhoods have been repeatedly attacked, with some completely destroyed (Roberts & Girke, 2024). The rapid urbanisation process in Yangon erases the traces of the past while deeply eroding urban memory.

3.3.3.5 Bangkok (Thailand)

Bangkok contains the spatial traces of political crises, military coups, and social uprisings within a complex urban fabric. During the Red Shirt protests in 2010, the fires and clashes in the Ratchaprasong area left permanent damage to shopping malls and surrounding buildings. In iconic areas such as Siam Square and Central World, buildings repaired after the fire partially conceal the traces of conflict, while bullet marks and damaged facades are still visible in some areas (King, 2017). In the historic neighbourhoods of Bangkok, such as Thonburi and Banglamphu, old temples and wooden houses have been displaced by unplanned urban redevelopment and infrastructure projects, disrupting the continuity of social memory (Ünaldi, 2014). Although the city's rapid modernisation attempts to erase the traces of past political violence and social trauma, the streets and public spaces of Bangkok continue to intertwine memory loss with the remnants of a traumatic past.

3.3.3.6 Hanoi (Vietnam)

Hanoi carries the spatial traces of urban amnesia and trauma after prolonged wars and occupations. In a city that was a target of American bombings during the Vietnam War, particularly in the neighbourhoods around Long Bien Bridge, bombed structures and repaired bridge supports showcase the physical remnants of war. Old French colonial buildings have either been abandoned or adapted for new uses in the post-war period; many bear marks of repairs and temporary interventions that make the trauma of the past visible (Johnson, 2002). In Hanoi's narrow streets, basements used as shelters during the war and old propaganda writings on walls serve as tangible evidence of social memory (Logan, 2016a, 2016b). The rapid economic transformation and modernisation in the city, along with the demolition of old neighbourhoods and the construction of new buildings, attempt to erase the traces of war, yet the traumatic memory of the past remains alive in Hanoi's urban landscape.

3.3.3.7 Seoul (South Korea)

Seoul embodies a complex interplay of urban amnesia and the spatial traces of war, influenced by the Korean War (1950–1953) and the subsequent rapid modernisation. The city, which was largely destroyed during the war, lost many of its old neighbourhoods during the reconstruction process, with traditional hanok houses in areas like Jongno and Myeongdong replaced by high-rise apartments (Lee, 2019). In the centre of Seoul, the Cheonggyecheon Stream was covered in the 1970s and reopened in the 2000s, becoming a natural public space; this process has produced a contradictory memory of both erasure and recollection of the past. The traces of war manifest not only in physical destruction but also in the psychological and social trauma created by the division between North and South (Youn & Ryoo, 2023). In neighbourhoods near the DMZ (Demilitarised Zone), shelters and military structures carry the spatial legacy of war into the present. Seoul continues to exist with an urban memory that oscillates between modernisation and forgetfulness.

3.3.3.8 Tehran (Iran)

Tehran clearly bears the traces of revolution, war, and social chaos in its urban fabric. The 1979 Iranian Revolution and the subsequent Iran-Iraq War (1980–1988) led to the destruction of many public spaces and buildings in the city. Particularly in southern Tehran, in industrial areas targeted during the war, bombed factories and repaired warehouse buildings still stand. Old cinemas in the city centre have either been closed or repurposed as a result of the ideological transformation following the revolution (Lak & Hakimian, 2019). On the walls of Tehran, images of war martyrs and revolutionary slogans are etched as traces of social memory. Rapidly rising skyscrapers and new residential areas replace the old neighbourhoods, while the traces of the past become increasingly faint (Golrokh & Saedi, 2024). Urban amnesia in Tehran manifests in both physical destruction and ideological reconstruction; the city seeks a balance between forgetting and remembering.

3.3.3.9 Beirut (Lebanon)

Beirut intertwines the spatial traces of civil war (1975–1990) and the subsequent political chaos with urban amnesia. The former "Green Line" in the city centre was once a war boundary separating East and West Beirut; today, the vacant lands, ruined buildings, and bullet marks still found along this line keep the spatial traces of conflict alive (Larkin, 2010). The rapid and intense reconstruction processes carried out in the post-war period, particularly through the Solidere project, have largely erased the identity and memory of the old neighbourhoods. Inns, markets, and hotels that were once central to social life have been transformed into luxury residences or commercial spaces. However, in areas like Hamra and Quarantine, abandoned buildings, bombed facades, and war graffiti persist as indelible traces of the past (Al-Harithy & Mneimneh, 2022). Urban amnesia in Beirut has opened deep fissures in the city's memory and spatial fabric as a result of both physical destruction and reconstruction.

3.3.3.10 Jakarta (Indonesia)

Jakarta carries the traces of colonialism, uprisings, and social chaos in a multi-layered urban fabric. During the anti-communist pogroms of 1965, many neighbourhoods in the city, particularly Glodok (Chinatown) and its surroundings, witnessed violent attacks and looting. In these areas, abandoned houses and repurposed temples reflect the erased layers of social memory (Kusno, 2010). During the Reformasi uprisings of 1998, shopping malls and business centres were set on fire, and many buildings still bear the marks of this period. The rapid urbanisation process in Jakarta, through the demolition of old neighbourhoods and colonial-era structures, seeks to erase the traces of the past while transforming the city's identity with new skyscrapers and shopping centres (Bijl, 2018; Wray, 2023). Urban memory is shaped by the traces of a traumatic past on one hand and the forgetfulness brought by modernisation on the other.

3.3.3.11 Bishkek (Kyrgyzstan)

Bishkek is an example of urban amnesia shaped by the political chaos, revolutions, and social upheavals of the post-Soviet era. During the popular uprisings of 2005 and 2010, government buildings around Ala-Too Square were attacked, and many structures suffered severe damage. The old Soviet monuments and public buildings surrounding the square have either been completely destroyed or repurposed with new ideological symbols (Harrowell, 2015). The old temples and Soviet-era apartment buildings in the city have fallen into ruin due to neglect and abandonment, while newly built mosques and shopping centres have accelerated the transformation of urban memory. The traces of revolutions and social conflicts can still be observed in the graffiti, crumbling walls, and empty buildings of Bishkek's streets (Flynn et al., 2014). The city serves as a stage for the ongoing struggle between the old and the new, memory and forgetfulness.

3.4 Discussion of Implications

In cities prone to conflict in Asia (Kabul, Lahore, Srinagar, Yangon, Bangkok, Hanoi, Seoul, Tehran, Beirut, Jakarta, Bishkek), urban amnesia is intertwined with the indelible marks left by war, occupation, internal conflicts, and social chaos on the spatial fabric of these cities. Although each city has experienced different historical and political processes, the collective memory of urban areas is systematically erased or suppressed through physical destruction and rapid modernisation processes. Bombed palaces, walls riddled with bullet holes, neighbourhoods turned to ruins, abandoned places of worship, and non-functional public buildings render the traumatic events of the past visible in today's urban landscape. On one hand, the rapid and often unplanned reconstruction processes following wars and conflicts replace old neighbourhoods and historical structures with shopping malls, skyscrapers, and new housing projects, breaking the continuity of urban memory and damaging social memory. On the other hand, military checkpoints, barricades, destroyed monuments, and war graffiti on walls continue to exist in the urban fabric as symbols of social trauma and a past that resist being forgotten. Alongside the rapid transformation brought about by modernisation, the identity and collective memory of these cities are simultaneously on the verge of being forgotten while resisting the traces of trauma. In these cities, urban amnesia signifies not only physical destruction but also a social forgetting, loss of identity, and a severing of ties with the past. These cities in Asia have become a stage for a continuous struggle between memory and forgetting, marked by the spatial traces of war and chaos.

The situation in Gaza clearly demonstrates that in 21st-century conflicts, there is neither sanctity nor adherence to any governing rules. It is apparent that occupiers in Gaza have indiscriminately targeted and destroyed places of worship, not only of a single religion but also those representing various faiths. Asia's conflict-prone cities (Kabul, Lahore, Srinagar, Yangon, Bangkok, Hanoi, Seoul, Tehran, Beirut, Jakarta and Bishkek) should, therefore, proactively establish detailed restoration projects and digital replicas of their immovable cultural heritage. These should be safeguarded in secure locations within both physical and digital realms in allied nations. Consequently, even if these invaluable cultural assets are damaged or destroyed in future conflicts, there will be a well-prepared blueprint for their reconstruction and preservation, ensuring the transmission of cultural heritage to future generations through multiple means. Gazans' persistent refurbishment and frequent repairs of their historical structures, which serve as sanctuaries during turbulent times, underscore the importance of these cultural edifices. Thus, it is prudent for conflict-prone cities in Asia to develop the areas surrounding their cultural assets as emergency sanctuaries and places of communal refuge during crises.

Asia's conflict-prone cities adjacent to water bodies (Yangon, Bangkok, Hanoi, Seoul, Beirut and Jakarta) must recognise port areas as critical strategic transportation hubs during potential conflicts. In Gaza, the bulk of human and material flow has relied on maritime routes rather than terrestrial ones. In times of crisis, these cities can maintain global connectivity through robust support from international humanitarian organisations via ports. Hence, fortifying port areas with military bases is essential to ensure resilience against potential blockades, attacks, or restrictions. Failing to do so could isolate a city and leave it to contend with myriad challenges on its own without international aid.

The phenomenon in Gaza has also demonstrated that, irrespective of severe conditions, traditional markets remain vital venues for fulfilling basic needs. Despite enduring tensions, restrictions, blockades, and economic sanctions, commerce persists in Gaza's traditional markets. Asia's conflict-prone cities (Kabul, Lahore, Srinagar, Yangon, Bangkok, Hanoi, Seoul, Tehran, Beirut, Jakarta and Bishkek) should anticipate that these markets will remain operational during adverse periods. Therefore, it is imperative to enhance their infrastructure and superstructure, ensuring they are self-sufficient and resilient for food, water, and energy logistics.

Conflicts and wars invariably intertwine sorrow and fleeting moments of joy. Gaza amnesia has revealed that central city squares can attract people for celebrations or protests amid chaos. Gaza's main square has served not only as a gathering space but also as a stage for global communication. In light of this, Asia's conflict-prone cities (Kabul, Lahore, Srinagar, Yangon, Bangkok, Hanoi, Seoul, Tehran, Beirut, Jakarta and Bishkek) should develop central squares with robust infrastructures capable of facilitating large gatherings under challenging conditions and consider incorporating multi-purpose shelters beneath these squares for emergencies.

Gaza amnesia has highlighted the critical role universities play in housing individuals capable of rational thought and judgment even during dire conditions. The Islamic University of Gaza's faculty and students have consistently and comprehensively communicated the ongoing tragedies to the world, employing evidence-based reporting. This transparency has allowed global awareness of Gaza's plight. Moreover, the university's resources have been pivotal during crises, serving the city's needs. Therefore, Asia's conflict-prone cities should design their universities as alternative storage and resource hubs, viewing these institutions not solely as educational centres but also as strategic assets capable of documenting and disseminating information during conflicts.

Among the most harrowing records of Gaza are those kept by Al-Shifa Hospital. Despite international legal protections, it has faced bombings, raids, and severe restrictions. For Asia's conflict-prone cities (Kabul, Srinagar, Hanoi, Tehran and Beirut), it is imperative to multiply and fortify health facilities to act as refuges during conflicts. Such enhancements should include making these facilities accessible via drones, helicopters, and air taxis, thus facilitating quicker medical aid delivery and patient transportation to safer locales. Additionally, equipping health facilities with energy production and storage systems will ensure continuous operation during grid failures, guaranteeing sustained medical care during crises.

Lastly, Gaza amnesia has underscored the vulnerability of beaches in wartime, revealing that wide-open areas offer little safety for civilians. Asia's conflict-prone cities must create and prepare designated safe zones for civilians, particularly women, children, the sick, and the elderly, to use for rest and social activities during challenging times. These areas should be cleverly designed subterranean spaces to provide safety from being overtly targeted while serving as collective shelters when necessary.

3.5 Future Directions

To prevent and mitigate urban amnesia in Asia, a multi-stakeholder and holistic collaboration model can be established among local governments, central governments, civil society organisations, academic institutions, and international organisations. Each city can develop concrete and applicable strategies that are suitable for its unique historical, cultural, and political context. For instance, in Kabul, the Darulaman Palace and its surroundings, as a symbolic area bearing the scars of war and destruction, can be preserved both physically and digitally through restoration and digital documentation projects; Afghan architects and historians can play an active role in this process, and support can be obtained from international cultural heritage funds. In Lahore, abandoned temples and caravanserais in the Old City can be transformed into functional cultural centres with the participation of local communities, thereby revitalising social memory and ensuring economic sustainability. In Srinagar, traditional houses and boats around Dal Lake can be restored with local artisans and youth, and long-term protection plans can be developed for structures that were demolished for security reasons, rather than temporary solutions. In Yangon, colonial-era buildings can be integrated into new commercial projects to prevent loss of identity, and independent cultural heritage foundations can be supported despite the pressures of military governance.

Additionally, crisis-resistant infrastructures can be developed according to the unique needs of cities. In Bangkok, in protest and conflict areas such as Ratchaprasong, multi-purpose shelters and community gathering spaces can be integrated underground, designing these areas as places where social memory is preserved and security is ensured during crises. In Hanoi, wartime shelters and propaganda walls can be transformed into city museums to be passed on to younger generations; legal protection status for old structures can be enhanced against the pressures of rapid urbanisation. In Seoul, financial incentives can be provided through public–private partnerships for the preservation of hanok neighbourhoods; in areas close to the DMZ, the physical remnants of war can be turned into parks for peace and reconciliation. In Tehran, old cinemas and post-revolution murals can be made accessible through digital archives and virtual tours; integrated designs that incorporate traces of the past can be encouraged in new constructions. In Beirut, the remaining ruins along the "Green Line" can serve to keep urban memory alive through memory walking routes and open-air exhibitions. In Jakarta, colonial and Chinatown structures in Glodok and its surroundings can be restored with the active participation of migrant communities, and references to this memory can be made mandatory in new skyscraper projects. In Bishkek, old Soviet buildings and revolutionary graffiti around Ala-Too Square can be transformed into urban memory projects with young artists and historians. All these proposals can be supported by transparent dialogue among stakeholders, sustainable financing, and social participation; a resilient, inclusive, and multi-layered memory policy can be adopted against urban amnesia.

3.6 Conclusion

This study delivers crucial findings about the future of Asian cities facing potential threats of war and conflict, derived from the urban amnesia observed in Gaza. It provides actionable recommendations, noting that Gaza's destruction and conflict have permanently damaged its historical and cultural fabric, leading to large gaps in its cultural memory. Consequently, it is imperative for Asian cities at risk to document and digitally archive their tangible and intangible cultural heritage while they can, and to prepare emergency plans for its protection. The turmoil in Gaza has severely compromised the city's infrastructure. Thus, it is prudent for Asian cities to fortify their infrastructure systems—such as water, electricity, and communication—making them resilient to conflict; to reinforce essential buildings like hospitals and schools; and to establish underground shelters and safe zones.

Cities experiencing destruction or occupation in Asia should explore technological solutions to transmit their collective memory, historical connections, and cultural identities to future generations through tangible evidence. Digital twin city technology emerges as an innovative tool that can significantly expedite the reconstruction process and preserve comprehensive city memories. Both portable and non-portable cultural assets provide the inspiration necessary for cities recovering from war and conflict to reconnect with their past and guide new administrators in identifying reconciliation points within society.

The situation in Gaza reveals a stark disregard for the urban norms, principles, and values collectively upheld in the United Nations Sustainable Development Goals. The ongoing urban crimes in Gaza represent a challenge to global democracy, human rights, and freedoms. Therefore, ending the occupation in Gaza—which has become a profound legal, moral, and ethical challenge for humanity—would foster hope for a peaceful future in Asia's conflict-prone cities.

International policies, global aid regulations, and transnational organisations possess the ability to formulate various approaches to alleviate the impact of urban amnesia. Specifically, these approaches can take decisive measures to safeguard and restore the cultural memory of cities affected by war and conflict. For instance, international bodies could establish funds and initiatives aimed at conserving cultural heritage that has been damaged or lost due to such events. By doing so, the physical losses contributing to urban amnesia can be averted, thereby ensuring the preservation of cities' identities and collective memories. Additionally, technological advancements serve as a crucial means for maintaining urban memory. International organisations could launch digital archiving projects to record lost sites and cultural assets following a disaster. Techniques such as creating digital twins of cities can assist local communities in rediscovering their history through alternative realities. To effectively combat urban amnesia, robust global policies are essential. Strengthening international human rights laws and penalising crimes against humanity and cities will aid post-conflict communities in their pursuit of justice.

The repeated destruction and reconstruction efforts in Gaza underscore the importance of urban planning. To enhance their resilience, Asia's conflict-prone cities should adopt flexible, adaptive urban planning strategies and leverage modular construction technologies for rapid rebuilding. These cities bear the responsibility, regardless of circumstances, to safeguard the elements that connect them to their land, culture, history, and identity (see Fig. 3.6 on the next page). Gaza's experience illustrates that wars target not only physical locations and people but also cultures and identities. Therefore, Asia's conflict-prone cities should devise methods to securely store some of their portable cultural assets in friendly nations.

Fig. 3.6 The first Ramadan iftar of Gaza's people in 2025, preserving their connection to land, culture, history, and identity. *Source* From the international online news portal, Mehrnews (2025)

Highlights

- Digital archiving and the preservation of cultural heritage with digital twins
- Strengthening infrastructure and critical structures, underground shelters, and emergency plans
- Enhancing cultural identity and resilience in Asia's conflict-prone cities through international cooperation and technology.

References

Abu-Amr, Z. (2015). The Gaza Economy: 1948–1984. In G. Abed (Eds.), *The palestinian economy-RLE economy of middle east* (1st ed., pp. 101–120). Routledge.

Adams, D. (2011). Everyday experiences of the modern city: Remembering the post-war reconstruction of Birmingham. *Planning Perspectives, 26*(2), 237–260.

Agaeva, N. (2021). Regeneration of the architectural heritage of cities based on urban ecology. *Spirit Time, 1*(1), 5–9.

Ahonen, P., Corni, G., Kochanowski, J., Schulze, R., Stark, T., & Stelzl-Marx, B. (2020). *People on the move: Forced population movements in Europe in the Second World War and its aftermath*. Routledge.

Al-Harithy, H., & Mneimneh, D. (2022). The [framing] of heritage in the post-war reconstruction of Beirut central district (Lebanon). In H. Al-Harithy (Eds.), *Urban recovery: Intersecting displacement with post war reconstruction* (1st ed., pp. 239–270). Routledge.

Almbaid, R. F. (2015). Gaza: The missing tourism assets. In R. K. Isaac, C. M. Hall, & F. Higgins-Desbiolles (Eds.), *The politics and power of tourism in Palestine* (1st ed., pp. 137–148). Routledge.

Alser, O., Asharaf, H. H., & Kishawi, S. (2024). Impact of the ongoing genocide in Gaza on surgical capacity and healthcare system: Gaza: Impact on surgical capacity. *Impact Surgery, 1*(3), 73–74.

Al Shobakib, M. J., Abu Amuna, Y. M., & Abu Naser, S. S. (2017). Organizational excellence in Palestinian Universities of Gaza Strip. *International Journal of Information Technology and Electrical Engineering, 6*(4), 20–30.

Arraf, H., & Shapiro, A. (2011). The free Gaza movement. In *Nonviolent resistance in the second Intifada: Activism and advocacy*. Palgrave Macmillan US.

As'ad, A. T. (2024). Exhibiting nation: A brief history of Palestinian exhibition making in the twentieth century. *Journal of Palestine Studies, 53*(1), 46–67.

Barreiro Leon, B. (2017). Urban theory in postmodern cities: Amnesiac spaces and ephemeral aesthetics. *URBS. Revista de Estudios Urbanos y Ciencias Sociales, Universidad de Almeria, 7*(1), 57–65.

Bevan, R. (2007) *The destruction of memory: Architecture at war*. Reaktion Books.

Bijl, P. (2018). Colonial memory and forgetting in the Netherlands and Indonesia. In B. Luttikhuis & A. D. Moses (Eds.), *Colonial counterinsurgency and mass violence* (1st ed., pp. 261–281). Routledge.

Bleibleh, S., & Awad, J. (2020). Preserving cultural heritage: Shifting paradigms in the face of war, occupation, and identity. *Journal of Cultural Heritage, 44*, 196–203.

Blockmans, W. P. (2003). Reshaping cities: The staging of political transformation. *Journal of Urban History, 30*(1), 7–20.

Burch, S., & Smith, D. J. (2007). Empty spaces and the value of symbols: Estonia's 'War of Monuments' from another angle. *Europe-Asia Studies, 59*(6), 913–936.

Cassar, B., & Noshadi, S. (2015). *Keeping history alive: Safeguarding cultural heritage in post-conflict Afghanistan*. UNESCO Publishing.

Cermeño, H., & Mielke, K. (2023). Endangered urban commons: Lahore's violent heritage management and prospects for reconciliation. *Urban Planning, 8*(1), 83–98.

Chang, H. (2020). Propositions for the aesthetic continuity of urban landscapes. *SAGE Open, 10*(3), 2158244020941851.

Crinson, M. (2005). *Urban memory: History and amnesia in the modern city*. Taylor & Francis.

Çalık, M., & Sözbilir, M. (2014). Parameters of content analysis. *Eğitim Ve Bilim, 39*(174), 33–38.

Diener, A. C., & Hagen, J. (2018). The city as palimpsest. *The city as power: Urban space, place, and national identity, 1*.

Dinçer, S. (2018). Content analysis in scientific research: Meta-analysis, meta-synthesis, and descriptive content analysis. *Bartın University Journal of Faculty of Education, 7*(1), 176–190.

Dogar, A. A., Shah, I., & Al Faisal, A. (2024). Cultural heritage under attack: Motives for deliberate destruction of cultural property in Armed Conflict in Swat. *Heritage & Society, 17*(2), 258–277.

Engbersen, G. (2001). The urban palimpsest. *Focaal-European Journal of Anthropology, 38*, 125–138.

Eren, F., & Yıldız Keçer, K. A. (2023). What is the contribution of a riverside design project to the establishment of social justice in the city? *Journal of Planning, 33*(1), 167–183.

Eren, Ş, Yorulmaz, N., & Eren, F. (2023). What can a school's corporate memory contribute to the school's students? *Milli Eğitim Dergisi, 52*(240), 2587–2614.

Falkingham, R. T., & Reeves, R. (1998). Context analysis: A technique for analysing research in a field, applied to literature on the management of R&D at the section level. *Scientometrics, 4*(2), 97–121.

Farrar, M. E. (2011). Amnesia, nostalgia, and the politics of place memory. *Political Research Quarterly, 64*(4), 723–735.

Feldman, I. (2007). *Governing Gaza: Bureaucracy, authority, and the work of rule, 1917–1967*. Duke University Press.

Flynn, M., Kosmarskaya, N., & Sabirova, G. (2014). The place of memory in understanding urban change in central Asia: The cities of Bishkek and Ferghana. *Europe-Asia Studies, 66*(9), 1501–1524.

Gilbert, M., & Fosse, E. (2009). Inside Gaza's Al-Shifa hospital. *The Lancet, 373*(9659), 200–202.

Goebel, S., & Keene, D. (2011). *Cities into battlefields: Metropolitan scenarios, experiences and commemorations of total war*. Ashgate Publishing.

Golrokh, S., & Saedi, M. (2024). Sites of disremembering: narrative tactics of remembering the past in Qasr and Heshmatieh neighborhoods of Tehran. *Iranian Studies, 57*(4), 534–559.

Gonzaga, E. (2019). Precarious nostalgia in the tropical smart city: Transmedia memory, urban informatics, and the Singapore golden jubilee. *Cultural Studies, 33*(1), 147–169.

Graham, S. (2008). *Cities, war, and terrorism: Towards an urban geopolitics*. Wiley.

Grodach, C. (2002). Reconstituting identity and history in post-war Mostar, Bosnia-Herzegovina. *City, 6*(1), 61–82.

Harrowell, E. (2015). From monuments to mahallas: Contrasting memories in the urban landscape of Osh, Kyrgyzstan. *Social & Cultural Geography, 16*(2), 203–225.

Hasan, M. (2019). Palestine's absent cities: Gender, memoricide and the silencing of urban Palestinian memory. *Journal of Holy Land and Palestine Studies, 18*(1), 1–20.

Hassouna, A., & Nam, K. C. (2005). Gaza Seaport as a Way to enhance the Palestinian economic development. *International Journal of Navigation and Port Research, 29*(9), 813–819.

Hobsbawm, E. J. (1994). *The age of extremes: The short twentieth century, 1914–1991*. Michael Joseph.

Huff, G. (2014) *Urbanization in Southeast Asia during the World War II: Japanese occupation and its aftermath*.

Hugo, G. (2006). Forced migration in Indonesia: Historical perspectives. *Asian and Pacific Migration Journal, 15*(1), 53–92.

İnce, G. B. (2014). Digital culture, new media and the transformation of collective memory. *Galatasaray Üniversitesi İletişim Dergisi, 21*, 9–29.

Johnson, M. (2002). Renovating Hue (Vietnam): Authenticating destruction, reconstructing authenticity. In R. Layton, P. Stone & J. Thomas (Eds.), *Destruction and conservation of cultural property* (1st ed., pp. 91–108). Routledge.

Junka, L. (2006). The politics of Gaza Beach: At the edge of the two intifadas. *Third Text, 20*(3–4), 417–428.

King, R. (2017). *Heritage and identity in contemporary Thailand: Memory, place and power*. NUS Press.

Klein, N. M. (2008). *The history of forgetting: Los Angeles and the erasure of memory*. Verso Books.

Kosciejew, M. (2024). The divergent fates of wartime Israeli and Gazan (Palestinian) cultural heritage: A preliminary inventory and conceptual analysis of Heritagization processes in GLAMs on separate sides of the Israel-Hamas War. *Heritage & Society,* 1–33.

Kusno, A. (2010). *The appearances of memory: Mnemonic practices of architecture and urban form in Indonesia.* Duke University Press.

Lak, A., & Hakimian, P. (2019). Collective memory and urban regeneration in urban spaces: Reproducing memories in Baharestan Square, city of Tehran, Iran. *City, Culture and Society, 18*, Article 100290.

Larkham, P. (2002). *Conservation and the city*. Routledge.

Larkin, C. (2010). Remaking Beirut: contesting memory, space, and the urban imaginary of Lebanese youth. *City & Community, 9*(4), 414–442.

Lee, H. K. (2019). *'Difficult Heritage' in nation building*. Springer International Publishing.

Lewis-McCoy, R. H., Warikoo, N., Matthews, S. A., & Foley, N. F. (2023). Resisting amnesia: Renewing and expanding the study of suburban inequality. *RSF: The Russell Sage Foundation Journal of the Social Sciences, 9*(1), 1–24.

Loftus, A. (2012). *Everyday environmentalism: Creating an urban political ecology*. University of Minnesota Press.

Logan, W. (2016a). Remembering wars in Vietnam: Commemoration, memorialisation and heritage management. In K. Reeves, G. Bird, L. James, B. Stichelbaut & J. Bourgeois (Eds.), *Battlefield events: Landscape, commemoration and heritage* (1st ed., pp. 237–254). Routledge.

Logan, W. (2016b). Whose heritage?: Conflicting narratives and top-down and bottom-up approaches to heritage management in Yangon, Myanmar. In S. Labadi & W. Logan (Eds.), *Urban heritage, development and sustainability* (1st ed., pp. 256–273). Routledge.

McDowell, S. (2016). Heritage, memory and identity. In P. Howard, & B. Graham (Eds.), *The Routledge research companion to heritage and identity* (1st ed., pp. 37–53). Routledge.

Meerow, S., Pajouhesh, P., & Miller, T. R. (2019). Social equity in urban resilience planning. *Local Environment, 24*(9), 793–808.

Mehrnews. (2025). Gazze'de ilk iftar sofrası kuruldu [The first iftar table was set in Gaza]. Retrieved March 2, 2025 from https://tr.mehrnews.com/news/1924898/Gazze-de-ilk-iftar-sofras%C4%B1-kuruldu-video

Milton, S., Elkahlout, G., & Barakat, S. (2023). Protecting higher education from attack in the Gaza Strip. *Compare: A Journal of Comparative and International Education, 53*(6), 1024–1042.

Mohammed Asqool, A., Ismail, S. M., & Nordin, R. (2023). The protection of children during armed conflicts: Israeli violations of international humanitarian law in three wars in Gaza. *UUM Journal of Legal Studies (UUMJLS), 14*(2), 535–556.

Musa, A., Crawley, J., Haj-Hassan, T., Inglis, R., & Maynard, N. (2023). Gaza, 9 years on: A humanitarian catastrophe. *The Lancet, 402*(10419), 2292–2293.

Nagle, J. (2017). Ghosts, memory, and the right to the divided city: Resisting amnesia in Beirut city centre. *Antipode, 49*(1), 149–168.

Palestinian Information Centre. (2023). War on Gaza—Targeting civilians, including school and university students and members of administrative and teaching staff, Devastating Impact on Educational Institutions amid Israel's declared war in Gaza Strip Statement. Retrieved March 2,

2025 from https://palast.ps/en/news/war-gaza-targeting-civilians-including-school-and-university-students-and-members?language_content_entity=en

Panjabi, S., & Winter, T. (2009). Understanding the tensions in place: Conflict and conservation in Kashmir. *Historic Environment, 22*(1), 19–25.

Qiu Sun, K. (2021). Overwriting the memory of a modern Ruin in Chile: From UNCTAD III to GAM. *ARENA Journal of Architectural Research, 6*(1), 6.

Rajagopalan, M. (2012). Preservation and modernity: Competing perspectives, contested histories and the question of authenticity. *The SAGE handbook of architectural theory* (pp. 308–324).

Rastouskaya, O. M. (2019). Phenomenon of urban memory: Social-philosophical analysis. *Proceedings of the National Academy of Sciences of Belarus, Humanitarian Series, 64*(1), 33–40.

Ristic, M., & Frank, S. F. (2019). *Urban heritage in divided cities*. Routledge.

Roberts, J. L., & Girke, F. (2024). Heritage as boundary object: The troubling potential of the Secretariat in Yangon, Myanmar. *International Journal of Heritage Studies, 30*(3), 365–381.

Saber, I. F. (2024). A 'cultural genocide': Which of Gaza's heritage sites have been destroyed?. Aljazeera. Retrieved January 14, 2024 from https://www.aljazeera.com/news/2024/1/14/a-cultural-genocide-which-of-gazas-heritage-sites-have-been-destroyed

Sargın, G. A. (2004). Displaced memories, or the architecture of forgetting and remembrance. *Environment and Planning D: Society and Space, 22*(5), 659–680.

Shehada, Z. M. (2020). The reflection of interreligious coexistence on the cultural morphology of the Grand Omari Mosque. *Journal of History Culture and Art Research, 9*(4), 146–163.

Silberman, N., & Purser, M. (2012). Collective memory as affirmation: People-centered cultural heritage in a digital age. In E. Giaccardi (Ed.), *Heritage and social media* (1st ed., pp. 13–29). Routledge.

Silver, H. (2010). Divided cities in the middle east. *City & Community, 9*(4), 345–357.

Smaling, D., Velsink, H., Groenveld, R., Booy, N. & Mol, A. (1996). *Gaza sea port*. Delft University of Technology Press.

Sugrue, T. J. (2017). *Making cities global: The transnational turn in urban history*. University of Pennsylvania Press.

Susser, I., & Schneider, J. (2003). Wounded cities: Destruction and reconstruction in a globalized world. In J. Schneider & I. Susser (Eds.), *Wounded cities* (1st ed., pp. 1–23). Routledge.

Tanpınar, A. H. (1961). *Saatleri Ayarlama Enstitüsü [Institute for the Adjustment of Clocks]*. Remzi Kitabevi.

Tavşancıl, E., & Aslan, E. (2001). *İçerik Analizi ve Uygulama Örnekleri [Content Analysis and Application Examples]*. Epsilon Yayınları.

Topal, A. (2012). Israeli blockade of Gaza and the attack on Mavi Marmara. *Milletlerarası Hukuk ve Milletlerarası Özel Hukuk Bülteni, 32*(1), 103–154.

TRT Haber. (2024). İsrail ordusu harabeye çevirdiği Şifa Hastanesi'nden çekildiğini doğruladı [The Israeli army confirmed its withdrawal from the Shifa Hospital, which it left in ruins]. Retrieved April 1, 2024 from https://www.trthaber.com/haber/dunya/israil-ordusu-harabeye-cevirdigi-sifa-hastanesinden-cekildigini-dogruladi-848117.html

Ulusal. (2022). İsrail'in katlettiği 17 çocuk unutulmadı [The 17 children killed by Israel have not been forgotten]. Retrieved June 12, 2025 from https://www.ulusal.com.tr/haber/11662756/israilin-katlettigi-17-cocuk-unutulmadi

Ünaldi, S. (2014). Politics and the city: Protest, memory, and contested space in Bangkok. In P. Liamputtong (Ed.), *Contemporary socio-cultural and political perspectives in Thailand* (1st ed., pp. 209–222). Springer Netherlands.

Verpoest, L., Engelen, L., Heynickx, R., Schmidt, J., Uyttenhove, P., & Verstraete, P. (2020). *Revival after the great war: Rebuild, remember, repair, reform*. Leuven University Press.

Wang, F., & Wang, F. (2016). Point space: Measurement of the urban memory of historic buildings. In *Beijing urban memory: Historic buildings and historic areas, central axes and City Walls* (pp. 11–85).

Wicke, C. (2023). Urban spaces. In Y. Gutman, J. Wüstenberg, I. Dekel, K. M. Murphy, B. Nienass, J. Wawrzyniak, & K. Whigham (Eds.), *The Routledge handbook of memory activism* (1st ed., pp. 229–236). Routledge.
Winter, T., & Panjabi, S. (2010). Leaving the buildings behind: Conflict, sovereignty and the values of heritage in Kashmir. In M. Langfield, W. Logan & M. N. Craith (Eds.), *Cultural diversity, heritage and human rights* (1st ed., pp. 259–273). Routledge.
Wray, J. (2023). National histories, private memories: Indonesia and the Japanese Occupation. In M. Budianta & S. Tiwon (Eds.), *Trajectories of memory: Excavating the past in Indonesia* (1st ed., pp. 99–116). Springer Nature Singapore.
Wyly, E. (2024). The death and life of great visual urbanism. *Visual Studies, 39*(1–2), 205–210.
Yandex Maps. (2024). Location of the Palestinian city of Gaza in the Asian continent. Retrieved April 18, 2024 from https://www.yandex.com/maps
Youn, H. C., & Ryoo, S. L. (2023). Conflicts and resolutions due to the expansion of urban heritage-focusing on historic sites and Hanok Areas in Seoul. *Architectural Research, 25*(2), 29–40.

Fatih Eren is an Associate Professor in the Department of Urban and Regional Planning at the Faculty of Architecture, Istanbul Technical University. He completed his Ph.D. on the internationalisation process of the Istanbul real estate market at the University of Sheffield between 2007 and 2013. He has more than 20 years of experience as an academician in the higher education sector and has been involved in many public and private sector supported research and development projects as a coordinator, manager, researcher and consultant. His areas of expertise are Urban and Regional Planning, Property Market, Real Estate, Platform Urbanism, and Digital Twin City.

Chapter 4
Urban Imaginary, Inclusive Planning, and the Future of Brand-Driven City Building: Satellite Cities in Vietnam vis-à-vis China, Japan, and India

Cluster of Future Studies: Technological Trends

Dang Dao Nguyen

Abstract This paper focuses on an emerging urban development strategy in Asian satellite cities. These cities are small and intermediate urban areas, located near a larger metropolitan centre, and often appearing in various forms like city gardens, new towns, eco-towns or new capital cities. By examining how urban imaginaries and inclusive planning shape the construction of a satellite province of Hanoi, Vietnam, Hung Yen from a comparative lens with China, Japan and India, this research aims to contribute to the knowledge production of Asian urbanism and new town development in particular across various Asian contexts, and extending Asian situated insights to theorising cities in the world. Regarding research methods, my research conducts content analysis of archival data, ethnography with forty in-depth interviews with government officials, local inhabitants in Hung Yen City, combined with existing literature on satellite city-building in other Asian countries. Examining small and medium-sized cities within this satellite province, namely Hung Yen City and Van Giang, reveals a disconnect between elite and public imaginaries of its green and smart future, alongside tensions between state-led, neoliberal urban planning and public desires for involvement, mirroring broader Asian satellite city trends in governance, branding, and infrastructural challenges. The case of Vietnam illustrates an inevitable trend in satellite city strategies across Asia, where city branding like eco-city or smart city will continue to shape urban regeneration in the future.

Keywords Satellite city · Urban imaginary · Inclusive planning · Urban future · Asia

D. D. Nguyen (✉)
Department of Architecture, National University of Singapore, Singapore, Singapore
e-mail: e1616139@u.nus.edu

R. Zainol et al. (eds.), *Future of Cities in Asia*,
https://doi.org/10.1007/978-3-032-01016-2_4

4.1 Introduction

In recent decades, there has been an ongoing discussion about how to understand cities in Asia and extend Asian-situated insights to theorising cities in the world. This paper focuses on a specific urban form that frequently results from this inter-referencing and neoliberal urban planning: satellite cities. In various Asian contexts, this relatively new development trend has appeared in diverse forms, such as city gardens, new towns, eco-towns, or new capital cities. In general, satellite cities refer to small and intermediate urban areas that are located near a larger metropolitan centre. This concept has become a popular option for Asian city-makers to tackle urban challenges, particularly overpopulation and housing shortages. Initially, new cities were primarily intended to provide an attractive residential alternative for middle-class citizens who sought refuge from densely populated cities. Since the mid-twentieth century, industrialised Asian countries like Japan, Korea, Singapore, Hong Kong, India, and China quickly adopted the concept of satellite cities to their urban systems, but not all of them have successfully relocated people from the large cities (Tokyo, Beijing, Shanghai, Seoul, and Delhi, for example) to their satellite cities. There are several reasons why they failed or succeeded, and the ways they dealt with the results of satellite city-building are also different based on their political and socio-economic dynamics.

Vietnam is no exception. Until 2024, there are currently ten satellite cities and provinces around Hanoi and Ho Chi Minh City, and the government has planned to build more in the next five years to promote even development across megacities and small and medium-sized cities. Hung Yen province is one of the key satellite provinces of Hanoi, which has been playing important role in the key economic triangle Hanoi—Hai Phong—Quang Ninh in northern Vietnam.

Focusing on the current development of two key cities in Hung Yen province, Hung Yen City (HYC) and Van Giang, I argue that although they have not been formally designated as Hanoi's satellites under the official planning plans, their geographical proximity and fast pace of urbanisation have position them as emerging potential satellite cities or peri-urban growth poles within the broader Hanoi metropolitan region. Their development trajectories and the mobility between the megacity, Hanoi and Hung Yen province reflect the growing spatial integration between Hanoi and its surrounding provinces and cities, driven by improved transportation infrastructure and growing housing markets catering to the metropolitan workforce. In nature, this is similar to the development of satellite cities in the world. This research, therefore, aims to contribute to the knowledge production of Asian urbanism and new town development in particular, from a comparative perspective. By focusing on two common conceptual aspects of a satellite city, urban imaginaries and inclusive planning, and situating the case of HYC in the largely top-down urban management context in Vietnam, I offer scrutiny of an off-the-map, third-tier satellite city in Vietnam, critically reflect on the state-of-the-art discussions in Asian urbanism, and highlight the future direction of urban development in Asia.

The paper seeks to answer how the state-led satellite city projects in Vietnam, both official or unofficial ones, have been intertwined and shaped by urban imaginaries from top-down perspectives of elites and experts, and the bottom-up understanding of ordinary citizens. Furthermore, it investigates how the (non)inclusive urban planning has affected and contributed to the current development and future trajectory towards a branded city in Vietnam and other Asian cities. In a broader context, I aim to understand why the concept of satellite cities has become increasingly popular in Asian urbanism and its implications for the future of cities.

Regarding research methods, my paper employs qualitative methods with data triangulation in ethnography during 2023–2024, combining in-depth interviews with 40 individuals from both elitist groups, including policymakers, city officials, experts, urban planners, and ordinary citizens and immigrants in the two unofficial satellite cities, HYC and Van Giang. Furthermore, I also conducted document analysis of official strategies, governmental reports, legal documents, news in the media, and secondary research papers to gain historical insights and background information on satellite cities in Asia to support the detailed explanations of the current situation and future of the city, and compare it with other cities in China, Japan, India, and Indonesia.

As for the paper's roadmap, an overview of satellite city development across different Asian countries will be examined. Following that, the literature review and analytical framework on satellite cities are utilised to analyse the current situation of Hung Yen City (HYC), and to lesser extent, Van Giang, the two major urban areas under Hung Yen province, as well as suggest some possible future developments for cities in Asia and around the globe. The findings show that even though satellite cities have become increasingly popular among city-makers across Asia, scholars like Shaw (1995) or Percival and Waley (2012) argued that there is no fixed or concrete set of factors explaining the success or failure of a satellite city, since not all satellite cities were designed to be self-contained. Nevertheless, there are some similarities that governments in Asia have learned from Western neoliberal planning, notably the strong role of the government in shaping the urban imaginaries of various groups and the role of inclusive planning to attract residents, businesses, and services, as well as reduce inequality in urban society (Dao, 2022). In the case of HYC, the Vietnamese government has more power to ensure the success of satellite city-building in the initial stage, yet it has been facing critical challenges that other satellite cities face, including asymmetrical power, social inequality, and token city planning and making. It also highlights an inevitable trend in satellite city strategies across the Global South, where city branding like eco-city or smart city will continue to shape urban regeneration in the future.

4.2 Literature Review

4.2.1 Satellite City Building Experience in Asia

Graham Taylor is largely credited with coining the term "satellite city" in academic literature in his book *Satellite Cities* (Taylor, 1915). In the late nineteenth century, concepts of 'new towns' emerged as a response to the overcrowded industrial cities of Europe and America by providing an attractive residential alternative for middle-class people who sought refuge from densely populated cities (Hall & Ward, 2014). Satellite cities then possess different spatial configurations and purposes, including gated communities, eco-cities, smart cities, innovation hubs, Special Economic Zones, and new capital cities (van Noorloos & Kloos-Terboer, 2018). Since the mid-twentieth century, Asian countries started building their own satellite cities based on the market-driven and neoliberal planning of the Global North, yet the development trajectories of satellite cities in each Asian country can be very different from others. Therefore, it is critical to explore the case of HYC in comparison to other Asian nations' experiences with satellite cities, particularly when they started building this type of city much earlier than Vietnam.

Japan was among the first Asian nations to regulate the expansion of Tokyo, its capital city, by adopting the National Capital Region Development Law in 1956. This law created a Green Belt around built-up areas in inner Tokyo to manage the expansion of urban areas. A decade later, the government decided to build several new towns, including Hachioji, Shin Yokohama, and Omiya, as well as sizable new regional employment hubs like Chiba's Makuhari Messe, Yokohama's Minato Mirai, and the new Tokyo Coastal Subcentre, which were all designated as satellite cities. This national spatial development strategy was a success because the population growth rate had considerably slowed down and the number of housing units in central Tokyo became increasingly sufficient throughout the late twentieth century (Glickman, 1977). Decentralisation of population for Seoul also took place in the late 1960s with the establishment of Seongnam city, a satellite city of Seoul, located 20 kms southeast of the capital. Unfortunately, the government failed to provide job opportunities for locals and immigrants, so a large number of them had to commute from Seongnam to Seoul daily, or moved back to Seoul. Learning from this failure, in the next two decades, in the Seoul Metropolitan Area throughout the 1980s, five fully functional, self-sufficient satellite cities were built, namely Bundang, Ilsan, Pyeongchon, Sanbon, and Joongdong. Until the end of the 1990s, nevertheless, such commercial towns still had not been able to attract big corporations to extend their operations and services to address the needs of inhabitants (Lee & Ahn, 2005).

In 1958, **China** announced the Great Leap Forward, a five-year economic plan that deemphasised large metropolitan areas and encouraged the growth of rural and small- to medium-sized cities. Cities and promote the development of small and medium-sized cities (Kamal-Chaoui, 2009). As a result, forty new satellite towns were developed by the Shanghai municipal government, notably Minghang, Wuging, Beiyangquiao, Nanchang, Huangdao, and Anting. This strategy was a success in alleviating its population burden (Fung, 1981). On the contrary, in **India**, building satellite towns has generally not been as successful. Shaw (1995) illustrated how six ring towns were constructed around New Delhi, while four new towns were built to serve as satellites of Hyderabad. Among them, Ghaziabad and Faridabad had strong industrial bases and an abundance of job opportunities for immigrants from Delhi. However, since the government failed to integrate land use with public transportation in urban planning, residents, particularly the poor, did not want to relocate to these new towns (Jain & Pallagst, 2015).

Although satellite cities have become increasingly popular among Asian city-makers, Shaw (1995) and Percival and Waley (2012) argued that there is no concrete set of fixed factors explaining the success or failure of a satellite city, since not all of them were designed to be self-contained. Since they were intended to be predominantly residential areas, many of them have proven to be effective in this regard. However, there are still some overarching commonalities. By offering an efficient alternative, these cities have been able to successfully restrain the expansion of larger municipalities. Strong financial assistance and policy support from the government also contribute significantly to decongesting populations in Hong Kong and Shanghai.

4.2.2 Emerging Themes of Satellite Cities

In the literature of urban geography, many scholars deliberately explore a theoretical framework to examine satellite cities as they are often embedded in the regional theoretical lens of world-class cities in Africa (Abubakar & Doan, 2017; Van Noorloos & Kloosterboer, 2018; Watson, 2014), and speculative urbanism in Asia (Bishop et al., 2003; Goldman, 2011; Percival & Waley, 2012). Van Leynseele and Bontje (2019)'s research is among the very few studies conceptualising the satellite city with four emerging themes, namely urban imaginaries, city-doubles, informality, and inclusive planning. While these themes are interrelated, not all satellite cities present all of them. As for city-regions and city-doubles, not all satellite cities are designed based on the desire to be part of a world-class league of large cities (Lowes, 2002). Furthermore, in terms of ordinary city development, they claimed that satellite cities were likely to adopt informal settlement practices from below and engage citizens in state-led policy reforms to shape the future of the city. However, in authoritarian nations like China and Vietnam, bottom-up city planning is either nonexistent or uncommon, and bottom-up interventions in many Global South countries are primarily reactive rather than proactive (Benedikter & Nguyen, 2018). Therefore, in the context of Vietnam, I purposefully emphasise two of the most relevant themes in the theoretical framework on satellite cities, namely urban imaginaries and inclusive planning, to generate a rich analysis and explore their implications for the future of the Asian and global landscapes.

4.2.2.1 Urban Imaginaries

Urban imaginaries, according to Lindner and Meissner (2018), are all over us, from postcard images of urban decay and nostalgia to new visions of smart, green, and resilient cities; from architectural depictions of hippie lifestyles to performative activism in pursuit of spatial justice; from metropolitan utopia in literature and film to urban realities in geolocation-based games and applications. These imaginaries have been significantly impacted by neoliberalism, global capital trends, and brandings like 'eco-city,' 'smart city,' or 'futuristic city' (Goldman, 2011). Since the 1990s, the historical intersection of the technology industry and world-city urbanism has led to the rise of global or world-class cities. European empires attempted to establish cities in their Asian colonies throughout the first 70 years of the twentieth century, such as Britain in New Delhi, Chandigarh, Shanghai, and Hong Kong; the Soviets in Central Asia; the French in Indochina; and Germany in Qingdao, which significantly influence on Asian city-building imaginaries and logics. For example, the construction of new capital cities like New Delhi, Jakarta, and Kuala Lumpur, were all employed European town planning and building models, or Singapore incorporated the "Dutch" model of satellite towns and the British "new town-garden city" utopianism (Hogan, 2018).

Furthermore, Douglass and Huang (2007), McGee (2010), Junghans et al. (2018), and Nguyen (2023a, 2023b, 2023c) all examine urban visions and their attributes in constructing city futures and urban design towards smart and green futures in Asia. Lindner and Meissner (2018) emphasise that it is crucial to study urban imaginaries to understand the future directions of the city because 'it is important to understand space as simultaneously material, conceptual, experienced, and practised.' (p. 2). Towards the end of the 2010s, satellite cities emerged as a useful instrument for executing strategies for economic growth, drawing in capital and business, and incentivising real estate to provide alternative housing options and invest in new towns across developing countries in Asia, particularly in Southeast Asia and South Asia (Percival & Waley, 2012).

To achieve these futuristic urban imaginaries in a short period of time to improve national urban competitiveness, satellite cities and other types of new towns are often turned into 'spaces of exception' where existing urban-related regulations and laws are not effectively enforced in exchange for accelerating the city-making process. Murray (2015) suggested that satellite cities often look forward to future techno-economic expansion and attempt to avoid any prior mistakes in urban planning, yet in real life, many countries in the Global South failed to reach their goals, resulting in modern urban dystopia or the emergence of ghost towns. These visions of the future are often generated by the imaginaries of several actors and communities in urban society; hence, they are fundamental to processes of conceptualisation, envisioning, and performing urban futures (Dunn, 2019).

In short, the understanding of the concept of satellite cities and other types of branded cities is primarily based on a Western norm of urban thought and design, resulting in marginalising a wide range of prospective visions, understandings, and practices of contemporary urbanity and hierarchising cities according to levels of development, as Robinson (2013) highlights.

4.2.2.2 Planning for Inclusive Cities

The government often provides land acquisition and infrastructure readiness to connect the new towns with existing urban hubs to support satellite city building, yet the role of the private sector in investment and providing jobs and services for the locals is also critical to the success of planned settlement (Jelili, 2012). Urban development policies have undergone a paradigm shift as national governments, international financial institutions, and numerous private companies adopt more neo-liberal public policies in response to the growing awareness of the urban crisis in Asia. Even in one-party states like Vietnam or China, the governments are taking a step back and allowing private capital to finance urban development projects in the form of public–private partnerships, shifting the relationship between the state and private sector in a more robust market economy. As a result, the adoption of neo-liberal policies and structural adjustment initiatives leads to the cut in public spending and reshapes state-centric forms of urban service delivery (Hogan et al., 2012).

Furthermore, Watson (2013) argues that satellite cities in the Global South are not only 'urban fantasies' that are financially beyond reach for poor citizens, but they also frequently result in the eviction and relocation of vulnerable individuals. Governments, experts, and public–private development coalitions often safeguard their decision-making position in city-making, which often results in gated communities in several Asian countries (Leisch, 2002; Breitung, 2012; Srinivas, 2015). Since these new settlements raise the prospect of growing disparities and marginalisation by primarily catering to the needs of the middle class and above, considering inclusive spatial planning is important when examining any satellite city.

Harms (2018) conceptualised the current situation in cities in the Global South as 'civility exclusion', or self-regulation and socio-cultural forms of exclusion. It refers to a variety of internalised, implicit, and frequently self-imposed behaviours and beliefs that regulate relationships, actions, and networks in Asian cities, which often encompasses 'civilised cities' in East and Southeast Asia, 'world-class cities' in South Asia, or 'worlding cities' (Ong, 2011). While they are frequently imposed from above, ideas about urban order, etiquette, modern dispositions, or rule-oriented behaviour typically coincide with the desires of mainstream groups in a city, namely middle-class citizens. In Asian countries, mainstream groups are often middle-class citizens, particularly in the case of new towns where the high relocation costs, expensive housing fees, and scarcity of blue-collar jobs often prevent the poorer members of society from moving to these cities (Liechty, 2003). These middle-class residents are often considered ideal residents of a satellite city as they prefer to 'act as a self-responsible and self-governing subject' (Zhang, 2018, 18), which can foster the process of constructing smart, green, civilised cities from scratch. However, focusing on mainstream, privileged groups might also contribute to the exclusion of low-class citizens who are poor and uncivilised. Chattopadhyay (2015) argued that examining inclusive planning requires certain considerations such as what stakeholders and actors are included in participatory planning frameworks, the decision-making and implementation, whether these arrangements enhance access to urban services for all urban dwellers, and if so, for what purpose. Despite the fact that these concerns have a profound impact on the circumstances and future prospects of a more inclusive city, few studies have really looked at them carefully in the context of overlooked, off-the-map cities in many Asian countries.

4.3 Methodology

My paper employs qualitative methods with data triangulation in ethnography. I spent sixteen weeks during 2023–2024 in HYC living with local people and interviewing forty individuals, including twenty-four local residents and immigrants living in HYC, mostly middle-class citizens; six government officers at the city and ward levels; six professionals in real estate and urban development companies; and six social workers working for civil society organisations. By talking, observing, and participating in their daily lives, my paper provides comprehensive, holistic insights into their opinions and behaviours regarding the ongoing construction of the satellite city there, as well as the nature and built environment of the place they live.

Regarding my positionality, since I am a Vietnamese researcher but not a local in HYC, I was able to easily reach out and talk to people, and many of them shared their honest and sometimes "aggressive and inappropriate" opinions about government policy. To ensure anonymity, their names in this paper have been changed. Furthermore, my paper will contextualise and compare the empirical findings from a Vietnamese city with existing research on other Asian countries to critically assess the broader patterns of satellite city development within the Global South.

To address two main research questions, namely:

(1) How has the development of (un)official satellite cities been shaped by urban imaginaries from the top-down (elites and experts) and bottom-up (ordinary citizens)?
(2) How has the (non)inclusive urban planning of HYC contributed to and been reflected in the current development and future trajectory towards a branded city of HYC in particular and similar Asian cities at large?

I structured my interview guide based on thematic analysis, urban imaginaries, inclusive planning, and the future of cities. The form of qualitative interviews is semi-structured, or 'conversations with a purpose.' This allows both the author and interviewees to explore unexpected patterns and insights that might arise during the interviews.

In addition to ethnography and in-depth interviews, I also collected data about satellite cities from various sources, including existing studies, policies, statistics, and reports from the government and NGOs, newspapers, surveys, interviews, and speeches that are relevant to the research questions, to conduct content analysis. Therefore, the paper uses data triangulation throughout the research. The next part will provide a detailed analysis of the complicated development of HYC.

4.4 Findings

4.4.1 An Overview: Situating Hung Yen in the Urban Landscape of Vietnam

In Vietnam, urbanisation has been a high priority on the national agenda since 1986, when political and economic reforms, such as the Doi Moi policy, were implemented (Chen & Dharmavaram, 2020). After more than 30 years of implementing the policy and transitioning to a socialist-oriented market economy, uneven development between different regions and cities has occurred, with a strong emphasis on the two largest cities, Hanoi and Ho Chi Minh City. In contrast, smaller and medium-sized cities, including HYC, have experienced a slowdown in growth due to a lack of investment and resources over the years. This phenomenon of being overlooked is not limited to Vietnam; several smaller cities in the Global South face similar challenges (Ruszczyk et al., 2020). Nugraha et al. (2023) describe these small and medium-sized cities as marginalised, underrepresented, off the radar, and undertheorised. Hence, examining the current state and future trajectories of these cities by refocusing attention on varied temporalities, relationalities, power relations, and socio-spatial configurations will offer more meaningful insights into urban studies.

In recent years, the Vietnamese government has started mobilising more resources to develop smaller and medium-sized cities in an effort to achieve a more equitable national urban development system. City-makers expect that satellite cities will motivate universities, hospitals, and entertainment facilities to relocate, thus reducing population congestion in the city centre. However, in reality, Hanoi's satellite cities have not been able to fulfil their intended functions, and the inner city remains overpopulated.

In this context, HYC, located to the southeast of Hanoi and serving as the core urban area of Hung Yen province, has risen in prominence within the national urban strategy. It is one of the 10 administrative units of Hung Yen province, with a population density of 1,364 people per km^2 (ranked 4th in Vietnam) (Hung Yen People's Committee, 2022). Due to its vital role as the administrative, economic, cultural, and social centre of the province, Hung Yen Province has been designated as one of the three key urban areas within the economic triangle of the southern Hanoi metropolitan region (Nguyen, 2023a, 2023b, 2023c). In particular, Decision No. 768/QD-TTg of the Prime Minister in 2016 approved the adjustment of construction planning for the Hanoi Capital region through 2030, with a vision extending to 2050, which includes Hanoi and nine surrounding provinces (Vietnam Prime Minister Office, 2016). As a gateway to the sea transport system (the Hanoi-Hai Duong-Hai

Phong corridor) and the economic triangle south of Hanoi (Hung Yen-Dong Van-Phu Ly), the national government expects the city to foster regional development, strengthen regional connectivity, and establish high-tech zones (Vo & Vo, 2023). At the local level, significant efforts have been made to invest in the transport infrastructure system to the west of the city, with 10 bridges planned across the Red River to improve connectivity between Hanoi, Hung Yen, and other provinces in the region. Newly developed areas to the west of the province, such as EcoPark and Vinhome Ocean Park, also play an important role in providing housing for residents moving from Hanoi.

However, despite policy support from both the national and local government levels, the current urbanisation rate of HYC remains relatively low, and its development indicators are still average compared to the rest of the Northern region. The total area of the province is 93,020 ha, of which the area designated for urban development is 10,602 ha, accounting for only 11.37%. Residential land in urban areas amounts to 3,495 ha, or 32.96% of the total urban land area. Consequently, official government discourses highlight the need for HYC to accelerate its socio-economic development quickly and comprehensively, capitalising on regional and national infrastructure opportunities to fulfil its potential as a driving force for the development of the Northern region and the future of the national urban system (Hung Yen People's Committee, 2022). The general direction for city planning is based on five key development principles, which include:

- Utilise the geographical advantages
- Attract investors to form large-scale, synchronous and modern urban areas
- Ensure sustainable and inclusive development, by linking economic development with social development and environmental protection
- Develop a green, smart city with synchronous infrastructure
- Ensure security, stability, social order and safety.

To realise this vision, the resolution of the 19th Party Congress of Hung Yen Province for the 2020–2025 term identified a focus on attracting investment through strategic planning. This includes prioritising large-scale projects, high-tech industries, supporting industries, digital technology sectors and transportation as key drivers to support the development of both Hanoi and Hung Yen.

4.4.2 Building Satellite Cites in Vietnam

The development of satellite cities in Vietnam is similar to other East and Southeast Asian countries. After the Doi Moi period, the turn to the market-driven economy of Vietnam has opened room for the private sector to cooperate with the state to finance urban projects. Satellite city planning in Vietnam is also advertised and coupled with the image of futuristic, smart, and green urban areas to attract new residents, particularly middle-class and wealthier citizens (Sorensen, 2016). In an interview with Le, a manager at the largest real estate company in Vietnam who has been in charge of urban projects in several suburban areas, he shared that public–private partnerships are an irreversible trend in new town and satellite city development. Real estate companies need to submit their ideas and implementation plans, which often include Western-led neoliberal and futuristic urban design like environmental sustainability and data-sharing smart technologies, to the local governments and relevant government agencies to win the bids. 'It often takes at least three years for them to develop such proposals, and their urban planning experts often refer to concepts of satellite cities or new towns in Asia like China, Singapore, and Korea, as these countries already started building their satellite city models much earlier than Vietnam.' Not only Le, but other architects, real estate brokers and urban developers I talked to also agreed that by learning and collaborating cross-border and cross-sector, they successfully promote the concepts, designs, and imaginaries of satellite cities among both the elite and ordinary citizens. HYC is one of the important components of the Hanoi Metropolitan Area to turn Hanoi into a world-class city and 'a globally interconnected metropolis by 2045,' said Deputy Secretary of the Hanoi Party Committee Nguyen Van Phong (Hanoitimes, 2023). Hence, although it is not officially stated as a satellite city of Hanoi, it is often viewed as one of them, particularly among experts and urban planners due to the proximity to Hanoi. Previously, the plan to relocate urban dwellers from the core to existing five satellite cities was not successful because there are very few housing projects, industrial parks, and other services for residents (Van, 2017). Furthermore, the urbanisation rate has happened at a very low rate, with agricultural land accounting for over 40%, which causes several challenges in terms of administration, legislation, and relocation compensation for the local people (Doan et al., 2023). However, until now, this strategy has been considered a failure, and the city-makers have been discussing altering the plan to other strategies like "city within the city" or reducing the number of satellite cities. Figure 4.1 presents the five satellite cities of Hanoi.

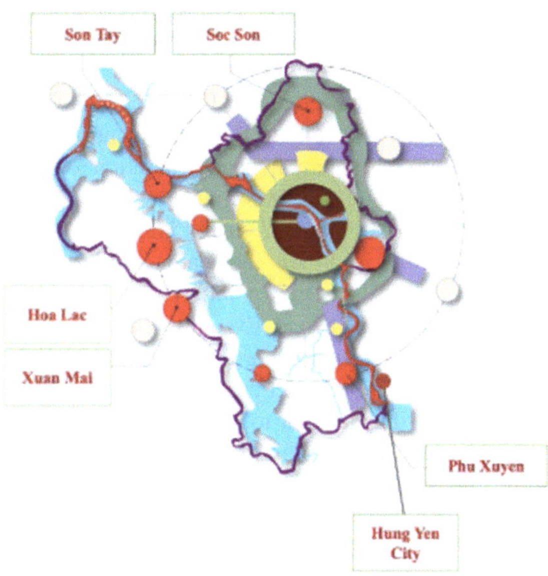

Fig. 4.1 Hanoi and its satellite cities. *Source* Hanoi Urban Planning Institute (2011). Plotted by the author

4.4.2.1 Urban Imaginaries of a Branded City

Hung Yen City (HYC) and Van Giang as emerging destinations for Hanoi residents has shown potential in lowering population density and housing demand for Hanoi with nearly 60 real estate projects during the period of 2021–2030 (Nguyen, 2023a, 2023b, 2023c). Hence, they have been viewing as satellite cities of Hanoi, particularly from the views of experts and real estate companies. Hoang Long, a government servant at one of the Ward People's Committees in HYC, shared his view on the housing landscape in Hung Yen during his interview with me: 'Nowadays, many individuals, especially young people and migrants living in Hanoi, prefer to reside in new towns and satellite cities like Hung Yen, Hoa Lac, and Son Tay. Hai Yen and her friends, young professionals working in Hoan Kiem district, a central district of Hanoi, all agreed that purchasing a home or apartment in any central district of Hanoi would take them at least 20 years of working. However, possessing a house ('an cư') is very significant in Vietnamese society, even before starting a family or working hard to advance in your job.' Therefore, my interviewees emphasised house ownership in HYC as the most important factor in why they decided to relocate to Van Giang and HYC as infrastructure across cities has been significantly improved. 'Vietnamese residents opt to live in satellite cities rather than take on massive debt in the hopes that as infrastructure improves, the distance from new urban areas to Hanoi will be greatly reduced.' HYC and Van Giang, in the imagination of many immigrants from Hanoi, are a 'cheap and convenient option for them to settle down.'

Moreover, the imagination of a future of HYC in the minds of the local residents is primarily shaped by state discourses and media since the media is largely controlled by the state. HYC residents shared that they 'only vaguely know about the masterplan and planning', 'hear about it when watching TV', or '... by talking with real estate brokers'. Hence, there are some patterns in what their imaginaries of the satellite city are and how they are shaped by the government. First, based on my in-depth interviews with 24 middle-class residents in different wards of HYC, most of them acknowledge that HYC and Van Giang are often unofficially viewed as a "supplementary" city to Hanoi. 'I spent over forty years residing in the Le Loi neighbourhood. From my father's generation to my son's, we frequently believe that the growth of Hanoi will be correlated with our own. HYC will be more prosperous if Hanoi is more developed,' shared Huy Viet, a middle-aged office man living in a newly built highrise in An Tao Ward. Similarly, Minh Ngoc, a recently graduated student, echoed Viet's viewpoint: "Everyone wants to relocate to Hanoi for their education and future career. Even if the cost of living in Hung Yen province is much lower, all the best career opportunities are in Hanoi. I may, however, get an apartment in a highrise in the adjunct area between two cities, like EcoPark and it will only take me more than an hour to commute between my house and my workplace if the transportation infrastructure is improved like the government promised.' Looking at the General Urban Development Strategy, it can be seen that what these citizens shared aligns with the strategy and the propaganda of the government, which is to 'leverage the geographical advantages of the province adjacent to Hanoi city.' They might not necessarily understand how a satellite city functioned, yet most of them said, “If other developed countries like Singapore, Japan, or Korea already used this model for their urban development, we should also follow them so our city will be modernised (‘hiện đại’) and civilised (‘văn minh’),” Nhan, the owner of a hotel I rented during my field trip, commented when we watched news about HYC on TV during dinner.

The concept of civility is very popular among Vietnamese cities and its neighbouring countries, where the state successfully engages citizens, mostly the middle class, to support their policies in order to move towards a more civilised society. As Erik Harms argued, ‘civility is an undeniably top-down dogma useful to a one-party state seeking to control a potentially unruly population.’ (Harms, 2014, 226). Indeed, the branding of smart ideas actually excites many inhabitants of the city because ‘it not only makes our lives more convenient, but it is also a proof of a modernised and civilised urban society.’ Trang, a supply chain manager at the global company, shared. This discourse on building civilised and modern cities is often the central discourse that is repeated many times, and affects the mindset, identity, and lifestyle of various groups in society. This is also where urban imaginaries come in and serve as the middle ground for the elites and ordinary people to co-develop ‘ideas to help guide the forces and complex situations of urban development.’ (Ache, 2017, 1).

Oftentimes the satellite city branding is coupled with other city brandings like ‘smart city’, ‘eco city’ or ‘green city’, as these Western-led concepts often associate with modernity and civility. The new branding coupled with the ‘satellite city’ title, namely the ‘smart city’ and ‘green city’ brandings, plays a key role in drawing the attention of investors and support from citizens for HYC. After Vietnam announced the strategy for smart urbanisation in 48 provinces and redistributed more resources and investment in technology infrastructure for low-tier cities, including satellite cities, that HYC and other small and intermediate cities receive more attention. One thing that Vietnamese city-makers often highlight in both official discourses and conversations with me is that they ‘always bear in mind and try to distribute resources equally between different regions and urban areas, even though there are many barriers constraining them from doing so, unlike some countries in Southeast Asia where resources are often concentrated in mega-cities like Metro Manila or Bangkok.’ This imaginary was highlighted in much of the above-mentioned research about satellite cities and other new towns in not only Asia but also other parts of the Global South, in which branded cities with advanced technological and sustainable features are considered the role models or benchmarks of ‘less developed’ countries and cities.

However, in the process of realising its smart city development goals, HYC also faces a similar situation as many satellite cities or new town across the world have had to tackle, which is the mismatch between the imaginaries and readiness, between vision and the existing conditions of the city. On the one hand, the ICT and transportation infrastructure and urban readiness of Hung Yen are not sufficient to reach the goals. Most city-makers acknowledge this shortcoming, as they pointed it out in our interviews and their reports to policymakers. Still, they largely share a positive view that the situation will be significantly changed as now the government has already adopted the strategy on smart cities: ‘The development of smart cities at the local level must be closely integrated with the digital transformation process, with the aim of being people-centred,' shared by Nguyen Huy Dung, Deputy Minister of Information and Communications (Vietnamnet Global, 2023). In the past 5 years, HYC has mobilised an average of 900 billion VND each year to invest in urban infrastructure and services. Until late 2022, the total number of investment projects in the city currently reaches nearly 100, with a total investment capital of nearly 9 trillion VND (Doan, 2022). After the province organised the Investment Attraction Conference in San Francisco, USA, in late 2023, urban developers are very optimistic about the future of the city, as they shared that 'to promote the growth of our city, HYC will undoubtedly draw in more investments from both domestic and foreign investors. We gave infrastructure development, especially technology development, first priority in order to assist high-tech application projects throughout the city and province.'

Apart from the autocratic imaginaries of the elites, citizens' understanding and awareness also pose significant challenges to the realisation of the smart and green city vision of HYC. Ha Linh, a secondary teacher, told me, 'I am excited to see my hometown become more modern, civilised, and green. We, as residents of HYC, must become more polite as the city becomes more modernised. However, I'm still uncertain about the "smart" part of the strategy. Does that imply a move towards greater automation and digitisation? I have never travelled abroad, so I can only saw modern cities in Chinese or American movies. I wholeheartedly support the government's goal if our city can be so modernised like other developed cities.' Other HYC residents who I spoke with gave me similar assurances that they would support the government in improving the quality of life: 'I believe it is for our common good and the government has been doing a great job'; 'I am happy they finally notice our city'; 'we have so much potential to be one of the excellent urban centres in the Northern region.'

Nevertheless, I acknowledge that the majority of my interviewees are working middle-class people who are beneficiaries of the government's urban development plan, although in official discourses and propaganda, the authorities at all levels need to consider engaging citizens in all stages of urban development. Doanh, a high-level officer at the city's Department of Construction, mentioned that not only him but his colleagues at other cities "did not consider inclusive planning and citizen participation are not a challenge, or at least an urgent one, among their daily tasks'. They 'always consult local residents several times before starting any projects and have to try so many tactics to address the needs of citizens to get their consensus on the tiniest matter in urban planning and construction'. In short, while there are some overlaps between different imaginaries of different groups of urban society, the mismatch between them is still evident, and the satellite city is being produced 'through collective action, negotiation, imagination, experimentation, and struggle.' (Ache, 2019, 2).

4.4.2.2 (Non)Inclusive City Making of Hung Yen, Vietnam

The question I want to answer regarding inclusive planning is whether public participation in urban development is only symbolic and tokenistic or meaningfully engaging. Either way, what is likely to be the future of a satellite city in Vietnam—a successful case like Shenzhen or ghost towns like Huairou or Baoding? In fact, Hung Yen province is well-known for the EcoPark in Van Giang, a leading eco-city model situated right next to HYC and paving the way for more eco-city projects across the province. It illustrates several issues regarding relocation packages for locals and exclusive urban planning in Vietnam. The investor in EcoPark, Viet Hung Corporation, referred to it as the 'perfect harmony of humans and nature.' (Ecopark, 2019). However, this urban fantasy of a green urban future is built at the expense of 4,000 families who were requested, or even forced to relocate and lost their livelihoods. Around 2,000 households refused to accept compensation during the land seizures and evictions that were carried out in 2009 and 2012 because, in their opinion, it was not sufficient for them to sustain their lives in a new location. It raises concerns for native residents of HYC when they hear about the strategy for building HYC as a satellite of Hanoi because 'many households are located in the satellite city planning or new projects zone, so we cannot build or renovate their houses. Therefore, we want the state to soon offer consolation and announce the relocation plan so that our families can move to a new place.' Uncle Duc, a retired professor, shared this with me.

Furthermore, several projects in the satellite city are often referred to as a 'gated community' for 'wealthy people who have the resources, time, and mindset to care about green, sustainability, and livability'. When I followed Huy, an urban planner in HYC, to work with local authorities and residents, he shared that within the urbanist community in Vietnam, they acknowledge how the gated community can further exaggerate social segregation and marginalise the poorer segment of society. Real estate companies still target the middle and upper class for housing projects in new towns because 'they are willing to pay more for the most advanced features in their house, like smart technologies or green indoor spaces, as these features will improve their quality of life. As for ordinary people like us, having a roof over our heads with basic services is enough; even though we love nature, it is not for the poor.' (see Fig. 4.2 on the next page).

This situation reflects the trend in many countries, where the government tries to relocate people from the centre to the periphery, where satellite cities are located, yet the price to buy a house and live there is out of reach for many people. That is why all of the ordinary people I interviewed all shared their strong desire for a more human-centred and inclusive approach from the government this time, where they will be consulted and even allowed to design and give feedback on the city-making process. Some residents recalled that the authorities sometimes come to their houses to ask for some details if they are directly related to the areas they are living in or send some surveys about access to urban services and provide feedback or complaints. However, they shared the same concern about whether their opinions have been actually considered and reflected in the implementation because they rarely hear

Fig. 4.2 A green building in Ecopark, Hung Yen, Vietnam. *Source* The photo was taken by the author during a field trip in January 2024

from governments or companies about how they incorporate specific concerns into the plans and execution.

Besides, some inhabitants living near the areas where the HYC authority planned to build the Pho Hien university areas complained that their lives have been unsettled because of the project for the past decade. 'I remember around ten years ago, an officer came to us and consulted us about a project to build new campuses at 10 top universities in our neighbourhood. My husband and I were so happy because our children can go to a university near our house, and we can also start a small business to sell daily products or rent services to university students. Since other satellite cities like Hoa Lac also successfully attracted universities and companies to relocate, we HYC residents were talking about it frequently, yet I only knew through the newspaper that the city suspended this project.' She also mentioned that rigorous civic consultations and a concrete action plan to address citizens' needs, both the local people and the migrants from megacities, are critical for a successful realisation of the urban imaginaries of the state and urban experts. This sharing from Phuong, a businesswoman who owns lots of land near this area, reminded me of the ghost town phenomenon across Asia.

Fig. 4.3 A ghost town in Baoding (a new town near central Beijing, China). *Source* The photo was taken by the author during a field trip in September 2023

The failure of the Pho Hien project signals negative future endeavours for satellite cities. The costs of high-rise and residential buildings built in ghost cities near major cities like Beijing and Shanghai are progressively rising as homeowners who do not plan to move there after buying them. Chau, a real estate expert, explained to me that HYC might face the same situation in the future because of a shortage of amenities necessary for daily life for urban dwellers. Without restaurants, schools, hospitals, and entertainment facilities, he predicted, "these places were unlikely to appeal to migrants from larger metropolitan centres, especially given that it was a huge investment, like what happened to the existing satellite cities of Hanoi." Figure 4.3 illustrates a ghost town in Beijing, China.

However, the Vietnamese government controlled land ownership, which created a path for Vietnam to mirror the initial success of satellite cities in China like Xiongan New Area (Hua, 2020). In **China**, since the central government has control over land, it was able to acquire land from local people to build infrastructure for the new city, assign real estate companies to provide housing for government officers, and select buildings for enterprises in information technology, biomedical, and new energy sectors to relocate to Xiongan New Area, keeping the price at a reasonable rate (Chiu, 2022). This is different from many Global North countries, where residents have the right to own their lands, and the government often struggles to negotiate and even forces people to relocate from the planned areas. In Vietnam, there is no effective mechanism to monitor land supply for urban projects, particularly the compensation plans provided by private companies. Huy, a land management expert, pointed out to me that in Vietnam, decisions on land supply for the market are decentralised to the local authorities and controlled by the monitoring and evaluation system operated independently and objectively by the central government. 'Although it might be convenient for the government and accelerate the construction process, it is an exclusive way of planning, which later might become a failed city, a city without humans.'

Furthermore, Vietnamese legality of urban land is unclear, and there are no separate regulations in the classification of current land law. The way of classifying land according to urban use purposes is not consistent with the practice of urbanisation. In light of that, when all Vietnamese citizens do not have the right to own the land and their suburban lands are subjected to the land conversion from agricultural to urban land to cater to satellite city projects, conflicts often arise. The compensation fee offered by the real estate and urban construction companies is often too low, which often leads to protests or opposition from local people. The situation in Van Giang and many other areas located in and near satellite cities proves that inclusive planning needs to be considered seriously and put highly on the agenda of city governance. Yet, in Vietnam, like other countries in the Global South, the state still remains ineffective and slow in improving its stakeholder engagement mechanisms, which will affect the future of the city in a negative way. At least in the foreseen future, HYC is likely to still witness the misalignment between the official planning and regulatory agencies' guidelines and the on-ground realities.

4.4.3 Compare Hung Yen City in Vietnam with Other Satellite Cities in Asia

The dynamics shaping Hung Yen City (HYC)'s growth reflect broader patterns of satellite city-building across Asia, revealing both similarities and divergences in urban imaginaries and realities. In **China**, the establishment of satellite cities such as Langfang, situated near Beijing, exemplifies a state-led strategy aimed at decentralising population and industrial activities from major urban hubs. This approach is facilitated by China's centralised governance framework, which effectively enables the rapid and efficient implementation of large-scale infrastructure initiatives. Langfang's integration of high-speed rail networks and the development of industrial zones have been instrumental in attracting both residents and businesses, thereby reinforcing its position as a complementary urban centre (Wu, 2016). Conversely, the case of Hung Yen province and its two cities, although have not been officially planned as satellite cities like other cases in Asia, still demonstrates a disconnect between its urban ambitions and the readiness of its infrastructure. The city's pursuit of "smart" urbanisation frequently exceeds its existing technological and transport capacities, highlighting a prevalent challenge in Vietnam's urban development pathway.

Similarly, as for the case of **Japan**, satellite cities such as Chiba and Saitama represent a well-established model of integration with their primary metropolis, Tokyo. These cities are supported by comprehensive public transportation systems, enabling efficient and seamless connectivity. Furthermore, Japan's focus on compact, high-density urban planning effectively curtails urban sprawl and fosters sustainable development (Sorensen, 2001). In contrast, HYC faces significant challenges in infrastructure development and connectivity, underscoring the disparity between its projected image as a "smart" and "green" city and the practical difficulties of realising such ambitions. Unlike Japan, where urban planning incorporates institutionalised mechanisms for public participation, HYC's planning processes remain predominantly top-down, offering minimal opportunities for meaningful citizen involvement beyond superficial consultations.

India provides a contrasting example, where satellite cities like Gurgaon (Gurugram) have evolved as centres of private-sector-driven urbanisation. Gurgaon's growth, propelled by corporate investments and speculative real estate activities, has resulted in the development of world-class infrastructure concentrated in select areas while neglecting equitable access to resources (Dupont, 2011). In contrast, urban development in HYC is predominantly shaped by the Vietnamese government, which strives to balance investment priorities with equitable growth. Nonetheless, HYC's branding as a "modern" and "civilised" city mirrors Gurgaon's emphasis on attracting middle-class residents and businesses, highlighting the pervasive influence of neoliberal urban imaginaries on global urban development narratives.

4.5 Discussions on Implications

Asia provides such a rich comparison framework to the Global North because it encompasses disparities that would typically be kept apart, such as megacities and global cities, "developed" and "developing" worlds, and cities from the Global South and North. The purpose of this paper, therefore, is not to generalise what the future of Asian cities looks like or should be. Rather, by unpacking the dynamics in HYC, I aim to provide another case study of a small city in a Global South country to the larger literature and knowledge production of Asian urbanism and new town development in particular. This purpose resonates with Douglass's (2009) call on relooking or reexamining the American-Euro-centric theoretical scope of Asian urbanism by both exploring how conformative they are in terms of expectations of a global or world-class city and how urbanists can produce other forms of urban development by extending Asian situated insights to theorising cities in other parts of the world.

The case of HYC also reveals interesting implications for satellite city planning, both official and unofficial, in the case of authoritarian and semi-authoritarian governance, which are the popular and on-the-rise governance types in Asia. Since the Vietnamese government has strong control over land management, real estate development, and propaganda, the process of gaining support from the local residents to develop a controlled and orderly satellite city has been smoother than in other highly unsuccessful cases in Japan, India, or China. Bhan (2014) described favourable conditions in (semi-)authoritarian states as free from the messiness of democratic politics. This might contribute to the rise of authoritarianism in Asia as other elite city-makers realise a more convenient path to develop new satellite cities to support the national agenda. Furthermore, a satellite city should be built with a multi-purpose character instead of only serving and catering to the specific needs of the larger city. In other words, to be self-contained and self-sufficient, a satellite city should be well-planned and multi-functional, where knowledge, technological, and environmental innovation are seamlessly integrated with residential, commercial, and speculative development. Abubakar and Doan (2017) also suggested that developing countries should avoid neoliberal, modernist planning, which often puts too much emphasis on urban fantasies. Instead, the central and city governments need to pay more attention to the local conditions by adopting more decentralised and inclusive planning to include the interests of various groups in society. Strengthening local competence and improving institutions to govern urban development is more effective than a centralised strategy and seeking international specialists who import expensive planning models, standards, and designs into low-tier cities in Asian countries like HYC without proper adaptations (Harrison, 2006).

4.6 Future Directions

The future of Asian cities will further highlight the importance of satellite cities and new towns in reducing the burdens on the most important megacities. However, to reciprocate success cases in industrialised countries, first, emerging economies like Vietnam and other Southeast and South Asian countries need to take into account the co-production of urban planning solutions with citizens that synchronise investments in technologies and sustainable built environments with sufficient local government resources. This is particularly relevant for the future direction of satellite cities since most of them are branded cities with smart, hi-tech, green, or futuristic models. City-makers also need to determine appropriate plans for urban renewal and expansion that optimise investment in sustainable and livable city features from various sources (UN ESCAP, 2022). Nevertheless, the prominent investment from real estate companies in HYC, which has been the source of major backlash and opposition from the citizens, proves that there is a strong need for the government in Vietnam and other countries in the region to diversify financial sources to build a more inclusive and sustainable future for Asian cities. Apart from investment from the private sector, potential sources of funding include private–public partnerships, loans, and investment from international investment banks like the World Bank, Asian Development Bank, or Asian Infrastructure Investment Bank, as well as state-to-state financial schemes.

Second, the examination of HYC supported Bunnell's (2012) claim that this kind of urban inter-referencing occurs not just on conventional North–South routes but also increasingly on South–South ones (Bunnell et al., 2012). City clusters will be the main drivers for global directions, and the satellite model, advocated by Chinese top leaders, will reflect in their investment in green development and the digital economy as a growth engine in Africa, Latin America, and even the Middle East (Wei, 2022). The development trajectory of cities in the Global South has illustrated the trend of building up an urban area from scratch as well-planned, sustainable, and self-contained communities outside of already-existing cities (Van Noorloos & Kloosterboer, 2018). Therefore, the current neoliberal and speculative urbanism approach in several Asian countries will likely continue to shape urban investment and building, particularly in the case of new town development in the Global South, to furnish world-class businesses and wealthier citizens, expats, and tourists. It means that satellite city development will keep being coupled with the concepts of technology, innovation, and sustainable hubs to attract wealthier sectors of urban society. While new towns like HYC are frequently constructed with the intention of fostering residential and commercial development, they are expected to serve as a national hub for technological innovation and development by drawing in universities, high-tech businesses, and start-ups while also facilitating local, national, and international connectivity. Whether they are designed to be an eco-city with urban experimentation of ecological innovation towards a low-carbon economy or a smart city in which its infrastructural foundation is operated based on continuous technology innovation, the reality and future of that vision, as Van Noorloos and Kloosterboer (2018) said, 'is still to be awaited.'

4.7 Conclusion

Satellite city, or new town development in general, is an urban expansion strategy that emerged as an option to improve the living conditions in cities by designing systematically planned communities that are organised, well-managed, and aesthetically pleasing. City-makers expect that satellite cities will motivate universities, hospitals, and entertainment facilities to relocate and reduce population congestion in the centre of Hanoi. Using the analytical framework of Leynseele and Bontje (2019), this paper analyses how the urban imaginaries of various actors and inclusive planning have shaped and will shape the future of HYC. The findings show that HYC is facing the same major issues as other peripheral provinces and cities located next to megacities. The development of satellite cities in Vietnam, although still at an early stage and often left unfinished, demonstrates some initial similarities to the process of satellite cities or new town development in East and Southeast Asia as a critical part of realising their imaginary of world-class cities. Satellite city planning in Vietnam is also advertised and coupled with the image of futuristic, smart, and green urban areas to attract new residents, particularly middle-class and wealthier citizens. Regarding urban imaginaries, since the planning for satellite cities includes multiple actors, there is always a mismatch between planning and reality and between urban elites and ordinary, marginalised residents, even though the imagination of a future of HYC in the minds of the local residents is primarily shaped by state discourses and media since the media is largely controlled by the state. Furthermore, the entrepreneurial and private-led nature of satellite cities often exacerbates social inequality and the poorer community of the city, which highlights the need for inclusive city planning and effective legal and land management mechanisms. Here, I want to extend the implication of urban imaginaries in studying Asian cities that Leynssele and Bontje had not mentioned. Theorisation of urbanism and the broader practices through which they travel, emerge from, and eventually form, imaginaries of cities (Burnell et al., 2012). By highlighting imaginaries of Asian cities, which in the neoliberal perspective, often share similarities with the rest of the developing world, challenge the stereotype of a typical Asian city—poor, overly crowded, unorganised, which have motivated city and central authorities to embark on competitive urban reimagining projects.

Despite some common traits, the HYC and Vietnamese urbanisation landscapes still have unique dynamics that contribute meaningfully to the scholarship on Asian urbanism and the future of cities. The imaginaries of the satellite city future of inhabitants in HYC are primarily shaped by the official discourses of modern and civilised urban dwellers, but the transition to a market economy and the increasingly important role of the private sector contribute significantly to the marginalisation of the relocation of inhabitants living in the planning areas of the satellite city. Hence, this study emphasises the lesson learned that satellite city building in Asia might be very different from the situation in the Global North, and there is no one-size-fits-all approach to all countries or cities. Applying the Western concept and method of building new towns might lead to massive failure for Asian states and other parts of the Global South in the future.

From a comparative perspective with other Asian satellite cities in India, China, and Japan, HYC reflects broader trends in Asian satellite city development, showcasing both similarities and divergences. In China, cities like Langfang demonstrate a state-led, centralised governance approach that efficiently decentralises population and industry, supported by advanced infrastructure. Japan's satellite cities, such as Chiba and Saitama, exemplify integrated urban planning with robust public transportation and institutionalised public participation, contrasting with HYC's infrastructural and participatory limitations. India's Gurgaon highlights private-sector-driven urbanisation, with unequal resource distribution, while HYC's state-led development emphasises equitable growth but mirrors Gurgaon's focus on attracting middle-class residents and businesses. Despite its ambitions as a "smart" and "green" city, HYC faces challenges in aligning its urban aspirations with infrastructural realities and citizen engagement.

The future of Asian cities will further highlight the importance of satellite cities in reducing the burdens on the most important megacities. However, to reciprocate success cases in industrialised countries, first, emerging economies like Vietnam and other Southeast and South Asian countries need to take into account the co-production of urban planning solutions with citizens that synchronise investments in technologies and sustainable built environments with sufficient local government resources. This is particularly relevant for the future direction of satellite cities since most of them are branded cities with smart, hi-tech, green, or futuristic models. Second, the examination of HYC supported Bunnell's (2012) claim that this kind of urban inter-referencing occurs not just on conventional North–South routes but also increasingly on South–South ones. Therefore, the current neoliberal and speculative urbanism approach in several Asian countries will likely continue to shape urban investment and building, particularly in the case of new town development in the Global South, to furnish world-class businesses and wealthier citizens, expats, and tourists. It means that satellite city development will keep being coupled with the concepts of technology, innovation, and sustainable hubs to attract wealthier sectors of urban society. However, the reality and future of that vision, as Van Noorloos and Kloosterboer (2018) said, 'is still to be awaited,' and their success in the future will require careful consideration of not only urban imaginaries and inclusive planning at the early stage, but also other factors like connecting infrastructure between the core and the satellites, identifying viable funding models and sources for continuing improvements in green and technological facilities, mitigating urban sprawl, improving the living standards of all residents, and achieving a sustainable future for Asian cities.

For future research on other new towns in the Global South, particularly in Vietnam, to provide more meaningful insights on the mismatch between the elites, experts, and policymakers and the ordinary people, it is suggested that human geographers, anthropologists, urbanists, and scholars in related fields need to pay more attention to the underprivileged groups, which are most affected by land appropriation and the lack of access and financial means to relocate to those towns. Furthermore, more future studies examining smaller cities like HYC are needed to enhance and broaden post- and de-colonial approaches that articulate understudied and underrepresented locations, knowledge, and practices in the Asian setting, shifting the primary emphasis of global urban studies to the urban majority.

Highlights

- Asian satellite cities show varied models, from China's state-led and Japan's integrated to India's private-driven and Vietnam's public–private models. However, most of them show tensions between aspiration and infrastructure, underscoring the uneven realities of smart, green urbanisation.
- The success of future satellite cities depends on balancing smart, green development with inclusive planning, funding, and equitable growth.

References

Abubakar, I. R., & Doan, P. L. (2017). Building new capital cities in Africa: Lessons for new satellite towns in developing countries. *African Studies, 76*(4), 546–565. https://doi.org/10.1080/00020184.2017.1370187

Ache, P. (2019). Vision making in large urban settings: Unleashing anticipation? In R. Poli (Ed.), *Handbook of anticipation: Theoretical and applied aspects of the use of future in decision making* (pp. 1327–1347). Springer. https://doi.org/10.1007/978-3-319-91563-0_51

Benedikter, S., & Nguyen, L. T. P. (2018). Obsessive planning in transitional Vietnam: Understanding rampant state planning and prospects of reform. *Journal of Vietnamese Studies, 13*(4), 1–47. https://doi.org/10.1525/vs.2018.13.4.1

Bhan, G. (2014). The real lives of urban fantasies. *Environment and Urbanization, 26*(1), 232–235. https://doi.org/10.1177/0956247813519350

Breitung, W. (2012). Enclave urbanism in China: Attitudes towards gated communities in Guangzhou. *Urban Geography, 33*(2), 278–294. https://doi.org/10.2747/0272-3638.33.2.278

Bishop, R., Phillips, J., & Yeo, W.-W. (Eds.). (2003). *Postcolonial urbanism: Southeast Asian cities and global processes*. Psychology Press.

Bunnell, T., Goh, D. P. S., Lai, C.-K., & Pow, C.-P. (2012). Introduction: Global urban frontiers? Asian cities in theory, practice and imagination. *Urban Studies, 49*(13), 2785–2793. https://doi.org/10.1177/0042098012452455

Chattopadhyay, S. (2015). Contesting inclusiveness: Policies, politics and processes of participatory urban governance in Indian cities. *Progress in Development Studies, 15*(1), 22–36. https://doi.org/10.1177/1464993414546974

Chen, M., & Dharmavaram, S. (2020). *Learning from Vietnam's urban upgrading projects*. Global Platform for Sustainable Cities. https://www.thegpsc.org/knowledge-products/urban-poverty-inclusive-cities-and-housing/learning-vietnam%E2%80%99s-urban-upgrading

Chiu, I. (2022). Cities lost in limbo: Are China's ghost cities here to stay? *US-China Today*. https://uschinatoday.org/features/2022/01/11/cities-lost-in-limbo-are-chinas-ghost-cities-here-to-stay/

Dao, N. D. (2022). Urban exclusion: Theoretical approaches and emerging trajectories for Vietnam. In A. T. Nguyen & L. Hens (Eds.), *Global changes and sustainable development in Asian emerging market economies* (Vol. 2, pp. 155–168). Springer. https://doi.org/10.1007/978-3-030-81443-4_8

Doan, L., Vo, H., & Pham, C. (2023). 5 satellite cities of Hanoi are ineffective for more than a decade [5 đô thị vệ tinh Hà Nội 'treo' hơn thập kỷ]. *VNExpress*. https://vnexpress.net/5-do-thi-ve-tinh-ha-noi-treo-hon-thap-ky-4637886.html

Douglass, M. (2009). Globopolis or cosmopolis?—alternative futures of city life in East Asia. *Studies in Urban Humanities, 2*, 67–115.

Douglass, M., & Huang, L. (2007). Globalizing the city in Southeast Asia: Utopia on the urban edge—The case of Phu My Hung, Saigon. *International Journal of Asia Pacific Studies, 3*(2), 1–42.

Doan, H. (2022). Hung Yen City aims to become a smart, ecological urban area [Thành phố Hưng Yên với mục tiêu trở thành đô thị thông minh, sinh thái]. *Hung Yen Department of Information and Communications*. https://sotttt.hungyen.gov.vn/portal/Pages/2022-7-4/Thanh-pho-Hung-Yen-voi-muc-tieu-tro-thanh-do-thi-tg8qt9m.aspx

Dunn, N. (2019). Urban imaginaries and the palimpsest of the future. In C. Lindner & M. Miriam (Eds.), *The Routledge companion to urban imaginaries* (pp. 373–386). Routledge.

Dupont, V. D. N. (2011). The dream of Delhi as a global city. *International Journal of Urban and Regional Research, 35*(3), 533–554. https://doi.org/10.1111/j.1468-2427.2010.01027.x

Ecopark. (2019). *Ecopark urban area—Where people blend with nature*. https://ecopark.com.vn/khu-do-thi-ecopark-noi-con-nguoi-hoa-quyen-voi-thien-nhien-n3203.html

Fung, K.-I. (1981). Satellite town development in the Shanghai city region. *The Town Planning Review, 52*(1), 26–46.

Glickman, N. (1977). *Growth and change in the Japanese urban system: The experience of the 1970s*. [Publisher not listed – please provide for complete citation if available].

Goldman, M. (2011). Speculative urbanism and the making of the next world city. *International Journal of Urban and Regional Research, 35*(3), 555–581. https://doi.org/10.1111/j.1468-2427.2010.01035.x

Hall, P., & Ward, C. (2014). *Sociable cities: The 21st-century reinvention of the garden city*. Routledge.

Hanoitimes. (2023). Hanoi aspires to become a global city. *Hanoitimes*. https://hanoitimes.vn/hanoi-aspires-to-become-a-global-city-324889.html
Hanoi Urban Planning Institute. (2011). Chapter VI: Spatial development orientation. In *Hanoi 903 capital construction master plan to 2030 and vision to 2050* (pp. 239–298).
Harms, E. (2014). Civility's footprint: Ethnographic conversations about urban civility and sustainability in Ho Chi Minh City. *SOJOURN: Journal of Social Issues in Southeast Asia, 29*(2), 223–262.
Harms, E. (2018). Marxist urbanism meets the specter of communism: Anthropological engagements with master-planned projects and mass dispossession in Ho Chi Minh city. In *The Routledge handbook of anthropology and the city* (pp. 297–309). Routledge.
Harrison, P. (2006). On the edge of reason: Planning and urban futures in Africa. *Urban Studies, 43*(2), 319–335.
Harrison, P. (2008). On the edge of reason: Planning and urban futures in Africa. In T. Harper, A. G. O. Yeh, & H. Costa (Eds.), *Dialogues in urban and regional planning* (pp. 213–236). Routledge.
Hogan, T. (2018). Twenty-first century Asian urban imaginaries and the *longue durée*. In T. Bunnell & D. Goh (Eds.), *Urban Asias: Essays on futurity past and present*. Jovis Verlag. [Add page numbers if available]
Hogan, T., Bunnell, T., Pow, C.-P., Permanasari, E., & Morshidi, S. (2012). Asian urbanisms and the privatisation of cities. *Cities, 29*(1), 59–63. https://doi.org/10.1016/j.cities.2011.07.005
Hua, X. (2020). Xinhua Headlines: Construction of Xiongan New Area back on fast track as epidemic eases. *Xinhua News*. http://www.xinhuanet.com/english/2020-03/31/c_138935111.htm
Hung Yen People's Committee. (2022). *Hung Yen Province plan general report during 2021–2030 with vision to 2050*. [Publisher not stated].
Jain, M., & Pallagst, K. (2015). Land use beyond control: How fragmented governance created sprawl in the Delhi Metropolitan Area. *disP—The Planning Review, 51*(3), 29–43. https://doi.org/10.1080/02513625.2015.1065418
Jelili, O. (2012). Urbanisation and future of cities in Africa: The emerging facts and challenges to planners. *Global Journal of Human Social Science, 12*(7), 7–12.
Junghans, L., Kreft, S., & Welp, M. (2018). Inclusive visions for urban transitions: Lessons from stakeholder dialogues in Asian medium-sized cities. *Sustainable Cities and Society, 42*, 512–520. https://doi.org/10.1016/j.scs.2018.07.014
Kamal-Chaoui, L., Leeman, E., & Zhang, R. (2009). Urban trends and policy in China. *OECD Regional Development Working Papers, 2009*(1). https://doi.org/10.1787/225205051036
Lee, C.-M., & Ahn, K.-H. (2005). Five new towns in the Seoul metropolitan area and their attractions in non-working trips: Implications on self-containment of new towns. *Habitat International, 29*(4), 647–666. https://doi.org/10.1016/j.habitatint.2004.04.004
Leisch, H. (2002). Gated communities in Indonesia. *Cities, 19*(5), 341–350. https://doi.org/10.1016/S0264-2751(02)00034-9
Liechty, M. (2003). *Suitably modern: Making middle-class culture in a new consumer society*. Princeton University Press.
Lindner, C., & Meissner, M. (Eds.). (2018). *The Routledge companion to urban imaginaries*. Routledge.
Lowes, M. D. (2002). *Indy dreams and urban nightmares: Speed merchants, spectacle, and the struggle over public space in the world-class city*. University of Toronto Press.
McGee, T. (2010). Building liveable cities in Asia in the twenty-first century: Research and policy challenges for the urban future of Asia. *Malaysian Journal of Environmental Management, 11*(1), 14–28.
Murray, M. J. (2015). City doubles: Re-urbanism in Africa. In K. Salo (Ed.), *Cities and inequalities in a global and neoliberal world* (pp. 92–109). Routledge.
Nguyen, D. D. (2023a). EU–ASEAN Smart Green ASEAN Cities Programme: Towards better urban connectivity in Southeast Asia. In *Dialogues on connectivity between Europe and Asia*

(pp. 155–168). Konrad-Adenauer-Stiftung e.V., Regional Programme Political Dialogue Asia, Diplomatic Academy of Vietnam, & Asian Vision Institute.

Nguyen, H. (2023b). Announcement of decision approving the partial adjustment of Hung Yen city master plan to 2035 [Công bố Quyết định phê duyệt Điều chỉnh cục bộ Quy hoạch chung thành phố Hưng Yên đến năm 2035]. *Hung Yen City Official Website*. http://thanhphohungyen.gov.vn/cong-bo-quyet-dinh-phe-duyet-dieu-chinh-cuc-bo-quy-hoach-chung-thanh-pho-hung-yen-den-nam-2035-c215920.html

Nguyen, L. (2023c). Hung Yen will have more real estate projects worth billions USD [Hưng Yên sắp có thêm nhiều dự án bất động sản tỷ USD]. *Vietnamnet*. https://vietnamnet.vn/hung-yen-sap-co-them-nhieu-du-an-bat-dong-san-ty-usd-2152279.html

Nugraha, E., Wesely, J., Ruszczyk, H. A., de Villiers, I., & Zhao, Y. (2023). Overlooked cities: Shifting the gaze in research and practice in global urban studies. *Cities, 133*, Article 104044. https://doi.org/10.1016/j.cities.2023.104044

Ong, A. (2011). Introduction: Worlding cities, or the art of being global. *UC Berkeley*. https://escholarship.org/uc/item/7xg0f2jx

Percival, T., & Waley, P. (2012). Articulating intra-Asian urbanism: The production of satellite cities in Phnom Penh. *Urban Studies, 49*(13), 2873–2888. https://doi.org/10.1177/0042098012453406

Ruszczyk, H. A., Nugraha, E., & de Villiers, I. (Eds.). (2020). *Overlooked cities: Power, politics, and knowledge beyond the urban South*. Routledge.

Robinson, J. (2013). *Ordinary cities: Between modernity and development*. Routledge.

Shaw, A. (1995). Satellite town development in Asia: The case of New Bombay, India. *Urban Geography, 16*(3), 254–271. https://doi.org/10.2747/0272-3638.16.3.254

Sorensen, A. (2001). Subcentres and satellite cities: Tokyo's twenty-first century planning challenges. *Urban Studies, 38*(12), 2283–2300. https://doi.org/10.1080/00420980120087294

Sorensen, A. (2016). Peri-urbanisation as the institutionalisation of place: The case of Japan. *Cities, 53*, 134–140. https://doi.org/10.1016/j.cities.2015.12.008

Srinivas, S. (2015). *A place for utopia: Urban designs from South Asia*. University of Washington Press.

Taylor, G. R. (1915). *Satellite cities: A study of industrial suburbs*. D. Appleton.

UN ESCAP. (2022). *The transition of Asian and Pacific cities to a sustainable future: Accelerating action for sustainable urbanisation*. https://www.unescap.org/resources/transition-asian-and-pacific-cities-sustainable-future

Van, H. (2017). Control population to reduce pressure on urban infrastructure [Kiểm soát dân số để giảm áp lực hạ tầng đô thị]. *Kinhtedothi.vn*. https://kinhtedothi.vn/kiem-soat-dan-so-de-giam-ap-luc-ha-tang-do-thi.html

Van Leynseele, Y., & Bontje, M. (2019). Visionary cities or spaces of uncertainty? Satellite cities and new towns in emerging economies. *International Planning Studies, 24*(3–4), 207–217. https://doi.org/10.1080/13563475.2019.1642567

Van Noorloos, F., & Kloosterboer, M. (2018). Africa's new cities: The contested future of urbanisation. *Urban Studies, 55*(6), 1223–1241. https://doi.org/10.1177/0042098017707960

Vietnam Prime Minister Office. (2016). Decision on the approval of adjustment of the construction planning of Hanoi city to 2030 and vision to 2050 [Quyết định về việc phê duyệt điều chỉnh quy hoạch xây dựng vùng thủ đô Hà Nội đến năm 2030 và tầm nhìn đến năm 2050]. https://vanban.chinhphu.vn/default.aspx?pageid=27160&docid=184573

Vietnamnet Global. (2023). Smart city development in VN faces challenges. *Vietnamnet*. https://vietnamnet.vn/hung-yen-sap-co-them-nhieu-du-an-bat-dong-san-ty-usd-2152279.html

Vo, H. T., & Vo, T. (2023). Vietnam's key economic zones: Performance and investments. *Vietnam Briefing*. https://www.vietnam-briefing.com/news/vietnam-key-economic-zones.html/

Watson, S. (2013). *City publics: The (dis) enchantments of urban encounters*. Routledge.

Watson, V. (2014). African urban fantasies: Dreams or nightmares? *Environment and Urbanization, 26*(1), 215–231. https://doi.org/10.1177/0956247813511184

Wei, A. C. (2022). New city clusters and satellite towns to be major boosters for China's economy. *The Straits Times*. https://www.straitstimes.com/asia/east-asia/new-city-clusters-and-satellite-towns-to-be-major-boosters-for-chinas-economy

Wu, F. (2016). China's emergent city-region governance: A new form of state spatial selectivity through state-orchestrated rescaling. *International Journal of Urban and Regional Research, 40*, 1134–1151. https://doi.org/10.1111/1468-2427.12437

Zhang, L. (2018). *In search of paradise: Middle-class living in a Chinese metropolis*. Cornell University Press.

Dang Dao Nguyen is a mid-career researcher in sustainable urban development, climate adaptation, and diplomacy. Prior to his doctoral studies at the National University of Singapore, he worked at the Vietnam Academy of Social Sciences, where he was awarded the Young Researcher of the Year 2023 award. He has also worked for the Cornell Center for Cities, the University of Oxford, Peking University, and the Konrad Adenauer Foundation on various research projects related to sustainability. He obtained his Master of Science in Global Affairs from Schwarzman College, Tsinghua University, and his Master of Law in China Studies from Peking University.

Part II
Future Direction of Urban Technology: Innovation and Smart Cities

Chapter 5
Digital Twin Revolution: Envisioning the Future of Transport Landscape in Bangkok, Thailand

Cluster of Future Studies: Past and Futures

Mahdi Aghaabbasi and Soheil Sabri

Abstract This chapter introduces a digital twin system designed to address the complex transportation challenges faced by Bangkok. As the capital of Thailand, Bangkok struggles with chronic traffic congestion, air pollution, and inefficiencies in its public transport infrastructure. The proposed system integrates real-time data collection, advanced simulations, and AI-powered analytics to create a dynamic virtual replica of the city's transport network. Key components include a data acquisition layer, a data integration platform, a simulation engine, an analytics module, and a visualisation interface. The framework supports applications such as traffic flow optimisation, public transport planning, predictive maintenance, and air quality monitoring. Lessons drawn from case studies in Singapore, Helsinki, and Amsterdam provide strategic insights for implementation. The chapter also considers future directions, including the integration with broader smart city initiatives and ethical considerations in data governance. While focused on Bangkok, this digital twin framework offers a scalable model for other Asian cities facing similar urban mobility challenges. It signifies a step towards a more connected, efficient, and sustainable urban future in Asia, positioning the region as a leader in shaping global smart city innovation.

Keywords Bangkok · Digital twins · Transport system · Real-time data · Transport network · Transport management · Transport planning

M. Aghaabbasi (✉) · S. Sabri
Urban Digital Twins Lab, School of Modeling, Simulation and Training (SMST), University of Central Florida, Orlando, USA
e-mail: mahdi.aghaabbasi@ucf.edu

S. Sabri
e-mail: soheil.sabri@ucf.edu

R. Zainol et al. (eds.), *Future of Cities in Asia*,
https://doi.org/10.1007/978-3-032-01016-2_5

5.1 Introduction

Bangkok, the vibrant capital of Thailand, is home to over 10 million residents and serves as the country's economic and cultural hub (World Population Review, 2024). Over the past decades, rapid urbanisation and population growth have placed considerable pressure on the city's transportation system, resulting in a wide range of complex and persistent mobility challenges. The city's multimodal transport infrastructure includes roadways, expressways, elevated toll ways, rail networks such as the BTS Skytrain and MRT subway (BTS MRTA, 2024; Skytrain, 2024), and water-based services along the Chao Phraya River. However, this network is marred by chronic traffic congestion, routinely ranking among the worst globally (Ayaragarnchanakul & Creutzig, 2022). Commuters frequently experience significant delays, especially during peak hours, while vehicular emissions contribute to dangerously high PM2.5 levels, posing serious public health risks (Wimolwattanapun et al., 2011; Jarernwong et al., 2023).

Public transportation in Bangkok faces several structural issues. Services are concentrated in central districts, leaving peripheral areas underserved. Accessibility remains limited, especially for elderly and disabled populations, due to inadequate pedestrian infrastructure and poor integration between transport modes (Permana & Petchsasithon, 2020; Chavanavesskul & Meksangsouy, 2022). The system also suffers from a heavy reliance on fixed-route buses, which contributes significantly to greenhouse gas emissions, particularly under conditions of severe traffic congestion (Champeecharoensuk et al., 2022).

Additionally, the lack of integration and interoperability across stakeholders undermines the effectiveness of Bangkok's Intelligent Transport System (ITS), leading to suboptimal operations (Choosakun & Yeom, 2021). Informal transport services such as tuk-tuks, motorcycle taxis, and Silor vans, while popular, often operate without adequate regulation, raising safety and operational concerns (Amrapala & Choocharukul, 2019; Nguyen et al., 2024). The COVID-19 pandemic further exposed vulnerabilities in ridership patterns, reinforcing the need for resilience in transport planning (Thaithatkul et al., 2023).

Efforts to address these issues have included infrastructure expansion (e.g., BTS and MRT), introduction of bus rapid transit (BRT) corridors, and traffic engineering strategies like reversible lanes and adaptive signal timing. Although policy proposals such as congestion pricing have been discussed, political resistance has hindered their implementation (Ayaragarnchanakul & Creutzig, 2022). The fragmented governance structure continues to impede cohesive planning and seamless passenger experiences, especially for vulnerable groups such as the elderly and disabled (Siangsuebchart et al., 2021; Permana & Petchsasithon, 2020).

Taken together, these challenges underscore the need for innovative, data-driven, and integrated approaches to reshape Bangkok's transportation landscape into a more sustainable, efficient, and inclusive system.

Digital twins are virtual replicas of actual objects, systems, or processes that employ real-time data to imitate, examine, and enhance their physical counterparts

(Attaran & Celik, 2023). A digital twin in urban planning refers to the creation of a dynamic and data-driven representation of a city or certain urban systems. This virtual depiction combines several data sources, including IoT sensors, satellite imagery, and demographic information, to construct a dynamic and comprehensive digital model of the urban environment (Deng et al., 2021a, 2021b).

The potential of digital twins in urban planning is substantial. Planners can utilise these tools to visually comprehend intricate urban systems, experiment with different situations, and forecast potential results without facing actual real-life repercussions (Ketzler et al., 2020; Nochta et al., 2021). For example, it is possible to simulate traffic movement in various scenarios, which can aid in optimising traffic signal timing or evaluating the effects of new infrastructure projects (Kumarasamy et al., 2023; Denazis et al., 2023). Utilising digital twins can help manage energy consumption by simulating building performance and finding areas for enhancing efficiency (Francisco et al., 2020).

Furthermore, these virtual models enhance decision-making by offering a comprehensive perspective on city operations. Urban planners have the ability to examine the connections between different urban systems, such as the impact of alterations in public transit on air quality or land use patterns (Holm et al., 2023). This interrelated approach facilitates the implementation of more cohesive and enduring plans for urban development.

Digital twins also improve public participation in urban planning procedures. City officials can enhance communication with communities and collect valuable feedback by developing immersive and interactive visualisations of proposed changes (Martone & Monica, 2024). This technology facilitates the connection between complex technical planning documents and the comprehension of the general population, promoting a more inclusive and participatory approach to urban development.

Within the realm of smart cities, digital twins function as a fundamental basis for governance that relies on data analysis. They facilitate the continuous monitoring of city performance in real-time, enabling the proactive repair of infrastructure and ensuring swift reaction to urban difficulties such as natural disasters or public health crises (Ford & Wolf, 2020). With the ongoing expansion of cities and the emergence of intricate problems, digital twins provide a potent solution for developing urban settings that are more robust, efficient, and enjoyable to live in.

Within the intricate transportation landscape of Bangkok, digital twins present a highly promising method for devising all-encompassing solutions. City planners could employ a virtual version of Bangkok's transportation system to test and analyze different scenarios to tackle the city's well-known traffic congestion. For example, a digital twin may simulate the effects of extending the BTS Skytrain or MRT subway to places that now lack sufficient service. It would forecast changes in traffic patterns and public transportation usage. It also has the potential to enhance the efficiency of the current road infrastructure by simulating various traffic signal timings or lane designs in order to enhance the flow of traffic. Additionally, a digital twin could help address Bangkok's air pollution issue by simulating car emissions in various traffic scenarios and evaluating the possible effects of measures such as low-emission zones or the adoption of electric vehicles. By creating virtual environments in which tuk-tuks and motorbike taxis interact with one another and with pedestrians, this technology has the potential to make Bangkok's many forms of public transportation safer and more efficient. Digital twins have the potential to assist Bangkok's planners in making well-informed decisions based on data by offering a platform to test and improve solutions before putting them into action. This might lead to a transformation of the city's transportation landscape.

5.1.1 Chapter Aim and Structure

The objective of this book chapter is to present a digital twin framework as a novel approach to tackle the complicated transportation issues in Bangkok. The goals of this study are to introduce the concept of digital twins and their possible applications in urban planning, analyse the existing transportation challenges in Bangkok, and examine the advantages and consequences of using this framework in the city's urban planning and transportation management.

The chapter will begin by offering an overview of Bangkok's transportation system and the notion of digital twins. Subsequently, it will explore an extensive overview of Bangkok's transportation infrastructure, followed by a comprehensive clarification of digital twin technology and its urban planning applications. The central focus of the chapter will be to introduce the proposed digital twin structure for Bangkok, encompassing its components, the implementation method, and possible applications. The discussion will encompass the advantages, challenges, and ethical deliberations of this method. The chapter will incorporate case studies of digital twin applications in other cities to provide context and insights, as well as analyse their relevance to Bangkok. Finally, it will examine the future ramifications of this technology in Bangkok and other urban areas, culminating in a concise overview of essential aspects and recommendations for future investigation and advancement. This framework aims to provide an extensive investigation into the use of digital twin technology to revolutionise Bangkok's transportation system and contribute to the wider discussion on innovative solutions for future cities.

5.2 Key Concepts and Issues

5.2.1 Digital Twin Framework

The proposed digital twin framework for Bangkok's transportation system includes the following essential elements (see Fig. 5.1 on the next page):

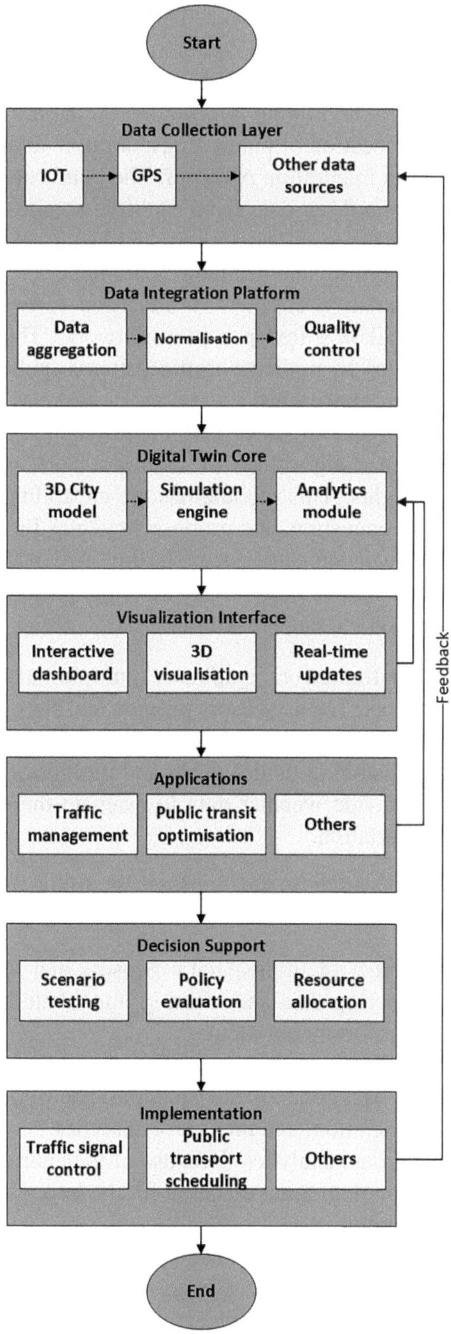

Fig. 5.1 The proposed digital twin framework for Bangkok's transportation system

- Data collection layer: This encompasses a system of Internet of Things (IoT) sensors, Global Positioning System (GPS) trackers, and cameras strategically placed throughout the city's transportation infrastructure. These devices gather up-to-the-minute information on the movement of traffic, the speed of vehicles, the location of public transportation, and the state of the environment.
- Data integration platform: The platform is a centralised system that collects and standardises data from multiple sources, such as traffic sensors, public transit systems, weather stations, and historical databases. This platform guarantees the uniformity and availability of data.
- Simulation engine: An advanced software system creates a virtual replica of Bangkok's transportation network. The system employs sophisticated algorithms to replicate traffic patterns, public transportation operations, and their interconnections.
- Analytics module: This module employs machine learning and artificial intelligence algorithms to scrutinise data, predict trends, and generate significant insights. This module has the capability to predict traffic patterns, detect areas of congestion, and propose strategies for improvement.
- Visualisation interface: The digital twin features an intuitive interface that displays data and simulations in a comprehensible way, enabling planners and decision-makers to engage with it.

The framework combines multiple data sources via APIs and data streaming protocols. Traffic sensors provide real-time data on vehicle movements and velocities. Transportation authorities directly receive data feeds to incorporate public transit data, including the locations and timings of buses and trains. Meteorological agencies provide weather data to evaluate the effects of environmental conditions on transportation.

This paradigm encompasses the following technologies (see Fig. 5.2 on the next page):

- Internet of things (IoT): Sensors and interconnected devices are essential for gathering data and offering immediate information on various elements of the transportation system.
- Artificial intelligence (AI) and machine learning (ML): These technologies are the backbone of the analytics and simulation modules' optimisation, pattern recognition, and predictive analytics.
- Big data analytics: Because of the enormous volume of data produced, big data technologies are essential for the large-scale storage, processing, and analysis of information.
- Cloud computing: The system offers the requisite computational prowess and storage capability to execute intricate simulations and manage vast datasets.
- 5G networks: Facilitate quick data transfer from Internet of Things (IoT) devices and instantaneous updates to the digital replica.

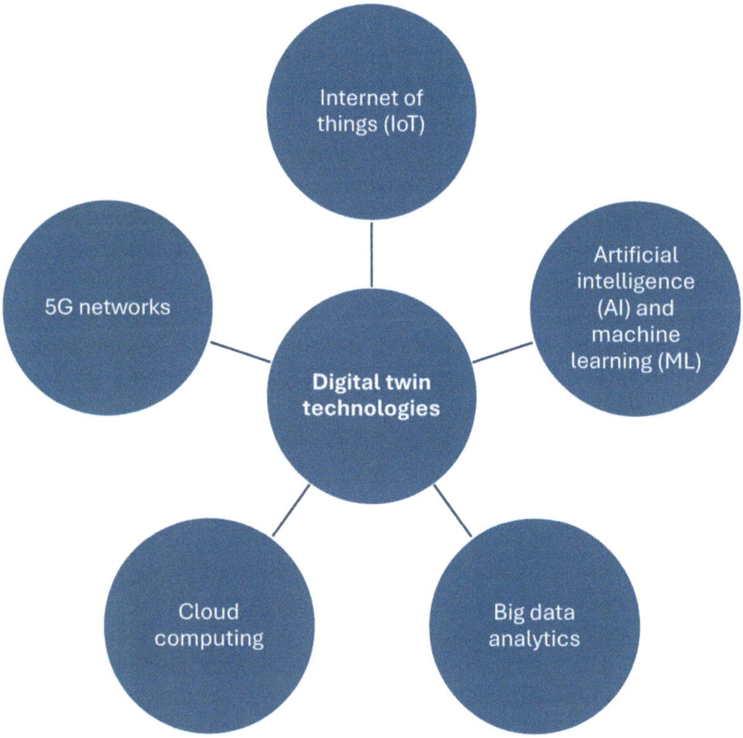

Fig. 5.2 Digital twin technologies proposed for implementation in Bangkok's urban digital twin system

5.2.2 Implementation

The process of implementing a digital twin for Bangkok's transportation system entails various crucial stages. To begin, it is necessary to create a thorough system for gathering data, which involves setting up Internet of Things (IoT) sensors, cameras, and GPS trackers across the city's transportation network. Next, a resilient data integration platform is created to collect and standardise data from several sources. The last phase entails constructing an elaborate 3D representation of Bangkok's transportation framework, encompassing roadways, railway networks, and water routes.

After establishing the basic components, the simulation engine must be created and adjusted, utilising both historical and real-time data. We then implement the analytics modules, integrating artificial intelligence and machine learning algorithms to generate valuable insights and forecasts. The last stage in the initial deployment process involves developing a user-friendly visualisation interface that allows stakeholders to interact with the digital twin.

During the implementation process, it is essential to actively involve different stakeholders, such as local officials, transit authorities, and technology providers. This partnership guarantees that the digital replica is in harmony with the requirements of the city and makes use of pre-existing systems and data sources.

Establishing this digital twin system presents a variety of potential challenges. The challenge of data quality and consistency is significant for the digital twin, as it depends on precise and up-to-date information from multiple sources. In order to tackle this issue, it is crucial to implement stringent data validation procedures and create regular maintenance methods for the sensors.

Privacy and security considerations provide a significant obstacle, especially considering the delicate nature of certain transportation data. To address these challenges, it is crucial to employ strong encryption techniques, enforce stringent access controls, and adhere to data protection standards.

The substantial expenses associated with implementing and maintaining the project provide a considerable obstacle, particularly for a large-scale endeavour in a metropolis such as Bangkok. This issue can be reduced by implementing it in stages, forming collaborations between the public and commercial sectors, and showing the long-term financial savings and improved effectiveness of the digital twin.

Technical obstacles include system integration and scalability. The digital twin must effectively interact with the current city systems and have the ability to manage larger amounts of data as the city expands. To resolve these problems, it is advisable to implement open standards and scalable cloud-based solutions.

Ultimately, there could be opposition to change from different individuals or groups with a vested interest in the matter. To overcome this challenge, it is necessary to implement a thorough change management plan that encompasses educational initiatives, showcasing first successes, and actively engaging stakeholders in the development process.

5.2.3 Global Case Studies Informing the Framework

Singapore developed Virtual Singapore, an extensive 3D digital twin platform, in 2014. It aims to generate a precise replica of the complete city-state (Erol et al., 2020; Hyun et al., 2024). The system combines live data, sophisticated analysis, and three-dimensional modelling to assist with urban planning and administration, fostering a sustainable city model (Eri & Elnæs, 2023). The platform has a range of uses, such as urban planning (visualising and analysing potential developments), infrastructure optimisation (determining the best location for cellular towers and solar panels), traffic management (simulating the movement of pedestrians and vehicles), disaster management, and cybersecurity (testing the city's defences against possible attacks) (Hyun et al., 2024; Erol et al., 2020; Vempati & Nalini, 2024; Potter & Olaoye, 2024). The efficacy of Virtual Singapore seems to depend on the seamless integration of up-to-the-minute data and advanced analytics. This implies that the successful implementation of a digital twin necessitates a strong data infrastructure, sophisticated data processing capabilities, and the capacity to evaluate extensive information. The platform also highlights the significance of citizen interaction, enabling feedback on matters related to urban planning and community development (Potter & Olaoye, 2024).

Helsinki's Digital Twin is a long-standing European project that has been developing for over thirty years. It has progressed from 2D CAD maps to a detailed 3D model that includes various city statistics (Du et al., 2024; Pauly & Pallagst, 2023). The digital twin depends on a strong data infrastructure that includes information such as energy use and traffic data (Pauly & Pallagst, 2023). The digital twin has a wide range of uses, such as urban planning, visualising different planning scenarios, conducting specific analyses, facilitating public participation, visualising data on mobility, analysing the potential of solar energy, optimising maintenance of infrastructure, and projecting urban heat islands (Du et al., 2024; Pauly & Pallagst, 2023). The implementation of Helsinki's digital twin has enhanced the city's ability to effectively address urban development difficulties, optimise urban planning procedures, and promote the integration of public responsibilities across many departments, leading to improved efficiency (Pauly & Pallagst, 2023). The project highlights the significance of strong data infrastructure in developing efficient urban digital replicas and showcases the potential influence of these tools on a city's digital modernisation (Pauly & Pallagst, 2023; Wajid et al., 2024).

Amsterdam's Smart City effort, which began in 2009, has more than 170 initiatives that have been collaboratively designed by people, government, and industry (Hovorushchenko et al., 2024; Khan & Kumar, 2023; Sao & Gupta, 2023). The program seeks to improve sustainability and safety while tackling urban issues such as traffic congestion, energy consumption, and public safety (Hovorushchenko et al., 2024; Sao & Gupta, 2023). Some important uses of this technology include implementing intelligent parking and traffic management systems, developing citizen engagement applications such as "Buur" and "Mobypark," installing smart energy meters, implementing flexible street lighting, and participating in programs like Experiments with Autonomous Transport (SMAPS) and the Sharing City project (Hovorushchenko et al., 2024; Khan & Kumar, 2023; Sao & Gupta, 2023). Key takeaways from Amsterdam's approach include the efficacy of involving citizens, the emphasis on pragmatic solutions, the significance of a grassroots approach, and the relevance of cooperation in the development of intelligent urban solutions (Hovorushchenko et al., 2024; Khan & Kumar, 2023; Sao & Gupta, 2023). This project showcases the possibility of integrating digital technology with data use across several industries. Integrating digital twin technology has the potential to improve data analysis, planning, and public participation in smart city programs, as shown in (Sao & Gupta, 2023).

These case studies provide significant insights for Bangkok. They emphasise the significance of inter-agency coordination, open data policies, citizen participation, gradual deployment, and integration with physical urban planning. By customising these lessons to its specific circumstances, Bangkok may optimise the efficiency of its digital twin implementation and maximise the advantages for its transportation system and overall urban growth.

5.3 Discussion of Implications

Optimising traffic flow is a crucial use of the digital twin framework for Bangkok's transportation system. The digital twin can utilise simulations of different traffic conditions to determine the most efficient signal timing, lane designs, and routing algorithms for minimising congestion. An example of its application is in simulating the effects of deploying adaptive traffic signal control systems, which dynamically adjust in response to real-time traffic circumstances (Kamal et al., 2024; Lim et al., 2023). The use of digital twins allows for testing different traffic management strategies by running parallel simulations, predicting their impacts without real-world risks. Digital twins improve situational awareness, provide insights into intricate dynamics, and facilitate the secure assessment of management tactics, indicating their potential for more streamlined and successful traffic management solutions (Kušić et al., 2023).

Digital twins have the potential to significantly enhance the efficiency of public transit systems, namely demand-responsive transit (DRT) (Deng et al., 2021a, 2021b). Digital twins have the ability to replicate real-world DRT operations by generating a digital model of certain trip sequences and bus operation sequences. This digital representation enables the calculation and assessment of the effectiveness of DRT services, including the evaluation of individual travel experiences such as waiting and riding times, as well as bus occupancy. Additionally, by examining data on the duration of service, frequency of runs, and passenger volume, digital twins can assess the level of bus service and its effectiveness. Digital twins have a dynamic quality that allows for the simulation of many situations in order to enhance route planning and resource allocation (Xiao et al., 2024). This includes evaluating factors like cost, emissions, and passenger demand. Furthermore, digital twins have the capability to adjust to immediate modifications and disturbances, guaranteeing dependable and effective public transportation services.

Digital twins greatly increase predictive maintenance efforts in transportation systems by enhancing dependability, efficiency, and safety. They provide comprehensive asset management services that cover the entire lifecycle, including specification, purchase, operation, performance evaluation, and disposal. The digital twin framework prioritises synchronising maintenance activities with business objectives in order to optimise profitability and maintain a high level of operational availability for transportation assets (Giel et al., 2023). This encompasses the formulation of a well-thought-out plan, the identification of essential information, the regulation of procedures, the implementation of tasks, and the ongoing enhancement of performance.

Regulation of air quality is an essential undertaking, particularly given Bangkok's pollution issues. Digital twins improve the management of air quality in cities by tackling the difficulties encountered by planners and decision-makers, especially in intricate metropolitan settings (Ariansyah et al., 2023). The system incorporates diverse data sources, including air quality indicators (PM2.5, NOx, and SO_2), meteorological data, traffic trends, and urban infrastructure information, to offer a thorough comprehension of air pollution dynamics. A digital twin system, comprising IoT sensors, smart city data sources, communication networks, models and analytics, user interfaces, and information displays, facilitates instantaneous interaction between the digital twin and the actual city. The digital twin is able to replicate the current air quality conditions through the use of real-time data collection and transmission. Visualisation capabilities can enhance decision-making by emphasising regions with low air quality and pinpointing variables that contribute to it, such as traffic congestion or industrial activity.

To illustrate the potential practical implementation of the proposed digital twin framework in Bangkok, let's consider a hypothetical scenario focusing on traffic congestion management in the central business district (CBD). In this proposed system, data collection would involve gathering real-time traffic flow information from a network of CCTV cameras and inductive loop detectors strategically placed throughout the CBD. Public transit data would be collected from GPS trackers on buses and trains, while environmental data would come from air quality sensors. The data processing stage would include cleaning, validation, integration, and normalisation, with edge computing nodes performing initial processing to reduce latency. Once processed, the data could be applied in various ways: for real-time traffic flow optimisation, the digital twin would analyse current patterns, simulate various scenarios, and recommend optimal traffic signal timings. For public transit optimisation, the system would analyse passenger load data to suggest route modifications or additional services. In terms of air quality management, the digital twin would correlate traffic congestion with air quality data, suggesting traffic diversion strategies when necessary. Urban planners could use the simulated data to visualise the impact of proposed infrastructure changes in the digital twin's 3D environment. While this scenario is hypothetical, it demonstrates how the proposed digital twin framework could potentially collect, process, and apply data to address Bangkok's complex urban transportation challenges, potentially resulting in improved travel times, increased public transit ridership, and better air quality. This example illustrates the practical applicability of the proposed framework and its potential to drive tangible improvements in urban mobility and quality of life in Bangkok.

5.4 Future Directions

As we look towards the future of digital twin technology in urban environments like Bangkok, it is crucial to consider the global context of this technological advancement. Global sharing of digital twin information is essential for advancing urban technology efficiently. Collaboration allows cities to avoid redundant efforts, reducing development costs and time. Notably, there is very limited information about digital twin project costs in both academic and non-academic sources, highlighting a critical knowledge gap. By sharing experiences, datasets, and methodologies, cities can enhance model accuracy, adapt proven strategies, and make more informed financial decisions. This cooperation accelerates innovation, promotes standardisation, and makes digital twin technologies more accessible to cities with limited resources. Ultimately, such information exchange fosters equitable and cost-effective paths towards smart, sustainable urban futures. For Bangkok, participating in this global exchange of digital twin information could prove invaluable in optimising the development and implementation of its own system, potentially reducing costs and accelerating the realisation of benefits.

The progression of Bangkok's digital twin throughout time is expected to be characterised by growing complexity and wider incorporation with other urban systems. With the advancement of artificial intelligence and machine learning technology, the digital twin has the potential to enhance its predictive powers, allowing it to foresee intricate urban phenomena with higher accuracy. The system may also gain increased autonomy, enabling it to make and execute specific decisions without human involvement, particularly in domains such as traffic management and public transportation scheduling.

The integration of the digital twin with other smart city programs offers substantial prospects for enhancing its capabilities and influence. For example, integrating the transportation digital twin with smart energy grids might enhance the efficiency of electric vehicle charging infrastructure and promote more effective energy utilisation throughout the city. Integrating with intelligent waste management systems could enhance the efficiency of garbage collection vehicles by optimising their routes, thereby minimising their disruption to traffic flow. With the increasing prevalence of Internet of Things (IoT) devices, the digital twin may now integrate data from a broader range of sources, resulting in a more thorough understanding of the urban environment.

The prospect of combining augmented reality (AR) and virtual reality (VR) technology with the digital twin is especially thrilling. These technologies have the potential to enable city planners and citizens to view planned alterations to the urban landscape in immersive 3D environments, thereby enhancing decision-making and public involvement. In addition, with the increasing prevalence of 5G and upcoming 6G networks, the digital twin will greatly improve its capacity to analyse and react to real-time data. This will result in more agile and efficient urban management.

Nevertheless, the development and growth of the digital twin also give rise to significant ethical considerations and concerns over data privacy. As the system accumulates and analyses additional data about individual citizens' activities and actions, there is a growing possibility of privacy violations and data abuse. It is essential to establish strong safeguards for data security, implement transparent policies for data governance, and set clear boundaries on data utilisation in order to uphold public confidence and safeguard individual rights.

The ethical considerations also encompass the decision-making powers of the digital twin. As the level of autonomy in the system increases, concerns arise regarding accountability and the possibility of algorithmic bias. Implementing supervision mechanisms and conducting frequent audits would be crucial to guaranteeing that the judgments and suggestions made by the digital twin are fair and equal for all citizens of Bangkok. Furthermore, as the digital twin becomes increasingly essential for urban administration, concerns regarding the digital gap and fair access to services may become more prominent. Measures should be implemented to ensure equitable distribution of the advantages of the digital twin among all sectors of society while also preventing any disadvantages for individuals lacking access to digital technologies.

In the future, the digital replica of Bangkok has the capacity to become an effective instrument for fostering a more sustainable, efficient, and habitable urban environment. However, harnessing this potential would necessitate careful oversight of technological advancement, deliberate incorporation with other urban systems, and a resolute dedication to ethical principles and societal fairness.

5.5 Conclusion

The suggested digital twin framework for Bangkok's transportation system embodies a complete strategy for tackling the intricate mobility issues faced by the city. The system combines real-time data collection, powerful simulation capabilities, and AI-driven analytics to construct a dynamic virtual duplicate of Bangkok's transportation network. The digital twin allows city planners and decision-makers to visually examine, analyse, and enhance many elements of the transportation system, including traffic flow and public transit operations.

The framework consists of several essential elements, including a solid foundation for collecting data, a centralised platform for integrating data, an advanced engine for simulating scenarios, and a user-friendly interface for visualising information. These parts collaborate to offer a comprehensive perspective of the city's transportation dynamics, allowing for better-informed and proactive urban management policies.

The potential influence of this digital twin framework on Bangkok's transportation system is substantial. The technology has the potential to improve traffic control, reduce congestion, improve public transportation efficiency, and improve air quality. The framework facilitates data-driven decision-making, leading to optimised resource allocation, enhanced emergency response planning, and bolstered sustainable urban growth. In addition, the digital twin's capacity to replicate different situations can significantly improve long-term planning endeavours, enabling city officials to evaluate and improve ideas prior to execution. This can result in more efficient regulations, more precisely directed infrastructure investments, and, ultimately, a more habitable and environmentally sustainable urban setting for Bangkok citizens.

Nevertheless, the effective execution and functioning of this digital twin structure would necessitate tackling many obstacles, such as data privacy apprehensions, technical integration complications, and the requirement for inter-agency cooperation. Successfully overcoming these obstacles will be essential in order to fully unlock the capabilities of the digital twin.

In the future, there are various areas that require additional investigation and advancement. The following items are included:

- Improving the forecasting powers of the digital twin using sophisticated machine learning techniques.
- Investigating the incorporation of cutting-edge technologies like 5G, edge computing, and augmented reality to enhance the capabilities of the digital twin.
- Creating strategies to guarantee fair and equal access to the advantages of the digital twin for all sectors of society.
- Exploring techniques to achieve a balance between thorough data gathering and strong privacy safeguards.
- Examining the long-term impacts of utilising digital twin-guided decision-making on the progress of urban development and the welfare of citizens.

To summarise, despite the existence of obstacles, the suggested digital twin framework has significant potential to revolutionise Bangkok's transportation system. By offering a robust instrument for comprehending, enhancing, and strategising urban mobility, it can make a substantial contribution to establishing a more effective, environmentally friendly, and enjoyable city for all inhabitants.

Highlights

- Digital twins offer scalable solutions for Asia's rapidly growing urban centres.
- Integration of AI, IoT, and real-time data enables predictive urban planning.
- Public–private partnerships and stakeholder engagement drive successful adoption.
- Ethical governance (privacy and equity) is essential for long-term public trust.
- Bangkok's model positions Asia as a global leader in smart city innovation.

References

Amrapala, C., & Choocharukul, K. (2019). Perceived service quality and commuter segmentation of informal public transport service in Bangkok, Thailand. *Engineering Journal, 23*(6), 1–18.

Ariansyah, D., Isnan, M., Rahutomo, R., & Pardamean, B. (2023). Digital twin (DT) smart city for air quality management. *Procedia Computer Science, 227*, 524–533.

Attaran, M., & Celik, B. G. (2023). Digital Twin: benefits, use cases, challenges, and opportunities. *Decision Analytics Journal, 6*, Article 100165.

Ayaragarnchanakul, E., & Creutzig, F. (2022). Bangkok's locked-in traffic jam: Price congestion or regulate parking? *Case Studies on Transport Policy, 10*(1), 365–378. https://doi.org/10.1016/j.cstp.2021.12.016

BTS Skytrain. (2024). *Company's profile.* Retrieved July 18, from https://www.bts.co.th/eng/info/info-history.html

Champeecharoensuk, T., Salam, P. A., Dhakal, S., & Chollacoop, N. (2022). Key driver analysis of greenhouse gas emissions in Thailand's public bus transport with comparative study on metropolitan Bangkok hotspots. *Energy for Sustainable Development, 70*, 456–465.

Chavanavesskul, S., & Meksangsouy, P. (2022). The ability to access attractions for the elderly using public transport in Bangkok metropolis. In *Applied Geography and Geoinformatics for Sustainable Development: Proceedings of ICGGS 2022* (pp. 247–258). Cham: Springer International Publishing.

Choosakun, A., & Yeom, C. (2021). Developing evaluation framework for intelligent transport system on public transportation in Bangkok metropolitan regions using fuzzy AHP. *Infrastructures, 6*(12), 182.

Denazis, S., Politi, T., Faliagka, E., Antonopoulos, C., Christopoulou, E., Tranoris, C., ... & Mihovska, A. D. (2023). Metacities excellence hub: Exploiting digital twins and metaverse technologies in south-eastern Europe. In *2023 IEEE International Smart Cities Conference (ISC2)* (pp. 1–5). IEEE.

Deng, S., Zhong, J., Chen, S., & He, Z. (2021a). Digital twin modeling for demand responsive transit. In *2021 IEEE 1st International Conference on Digital Twins and Parallel Intelligence (DTPI)* (pp. 410–413). IEEE.

Deng, T., Zhang, K., & Shen, Z. J. M. (2021b). A systematic review of a digital twin city: A new pattern of urban governance toward smart cities. *Journal of Management Science and Engineering, 6*(2), 125–134.

Du, Z. Y., Wang, Q., & Yu, Y. (2024). Asymmetric spatial competition and dynamics in digitalization: Will public R&D support and marketization exacerbate spatial polarization? *Cities, 152*, Article 105224.

Eri, A. B., & Elnæs, I. A. (2023). *The emergence of digital twin technology in urban planning: A study of perceptions, opportunities and barriers.* Master's thesis, Norwegian University of Life Sciences.

Erol, T., Mendi, A. F., & Doğan, D. (2020). Digital transformation revolution with digital twin technology. In *2020 4th International Symposium on Multidisciplinary Studies and Innovative Technologies (ISMSIT)* (pp. 1–7). IEEE.

Ford, D. N., & Wolf, C. M. (2020). Smart cities with digital twin systems for disaster management. *Journal of Management in Engineering, 36*(4), 04020027.

Francisco, A., Mohammadi, N., & Taylor, J. E. (2020). Smart city digital twin–enabled energy management: Toward real-time urban building energy benchmarking. *Journal of Management in Engineering, 36*(2), 04019045.

Giel, R., Werbińska-Wojciechowska, S., & Winiarska, K. (2023). Employing digital twins in operation and maintenance management of transportation systems. In *International Conference TRANSBALTICA: Transportation Science and Technology* (pp. 67–76). Cham: Springer Nature Switzerland.

Holm, J., Mohammad, P., Dawn, C., Kabir, N., & Pratyush M. (2023). *Creating a digital twin for transportation and air quality.* Los Angeles: City of Los Angeles.

Hovorushchenko, T., Baranovskyi, V., Ivanov, O., & Hnatchuk, A. (2024). Subsystem for monitoring atmospheric air quality in the cyber-physical system smart city. *Computer Systems and Information Technologies, 1*, 17–26.

Hyun, C. T., Lee, S., & Jin, C. (2024). Digital twin for disaster management in smart city. In *Future of Information and Communication Conference* (pp. 627–641). Cham: Springer Nature Switzerland.

Jarernwong, K., Gheewala, S. H., & Sampattagul, S. (2023). Health impact related to ambient particulate matter exposure as a spatial health risk map case study in Chiang Mai, Thailand. *Atmosphere, 14*(2), 261.

Kamal, H., Yánez, W., Hassan, S., & Sobhy, D. (2024). Digital-twin-based deep reinforcement learning approach for adaptive traffic signal control. *IEEE Internet of Things Journal.*

Ketzler, B., Naserentin, V., Latino, F., Zangelidis, C., Thuvander, L., & Logg, A. (2020). Digital twins for cities: A state of the art review. *Built Environment, 46*(4), 547–573.

Khan, Z., & Kumar, R. (2023). Building smart cities: Integrating IoT technologies for success. *Journal of Nonlinear Analysis and Optimization, 14*(1), 49–55.

Kumarasamy, V. K., Saroj, A. J., Liang, Y., Wu, D., Hunter, M. P., Guin, A., & Sartipi, M. (2023). Traffic signal optimization by integrating reinforcement learning and digital twins. In *2023 IEEE Smart World Congress (SWC)* (pp. 1–8). IEEE.

Kušić, K., Schumann, R., & Ivanjko, E. (2023). A digital twin in transportation: Real-time synergy of traffic data streams and simulation for virtualising motorway dynamics. *Advanced Engineering Informatics, 55*, Article 101858.

Lim, H., Go, M., Lim, T., & Kim, D. Y. (2023). Adaptive traffic signal control for a mixed autonomous and traditional vehicles by agent-based digital twin simulation. In *IFIP International Conference on Advances in Production Management Systems* (pp. 603–618). Cham: Springer Nature Switzerland.

Martone, A., & Monica, B. (2024). *Digital twin for urban development*. Cham.

MRTA. (2024). *About MRTA*. Retrieved July, 18,from. https://www.mrta.co.th/en/about-mrta-know

Nguyen, S. T., Moeinaddini, M., Saadi, I., & Cools, M. (2024). Unveiling the drivers of modal switch from motorcycles to public transport in Southeast Asia. *Transportation Research Part F: Traffic Psychology and Behaviour, 102*, 373–391. https://doi.org/10.1016/j.trf.2024.03.009

Nochta, T., Wan, L., Schooling, J. M., & Parlikad, A. K. (2021). A socio-technical perspective on urban analytics: The case of city-scale digital twins. *Journal of Urban Technology, 28*(1–2), 263–287.

Pauly, J., & Pallagst, K. (2023). The potentials of digital tools to contribute to spatial transformations—the example of the digital twin of the city of Helsinki. In *LET IT GROW, LET US PLAN, LET IT GROW. Nature-based Solutions for Sustainable Resilient Smart Green and Blue Cities. Proceedings of REAL CORP 2023, 28th International Conference on Urban Development, Regional Planning and Information Society* (pp. 763–775). CORP–Competence Center of Urban and Regional Planning.

Permana, A. S., & Petchsasithon, A. (2020). Assessing the seamlessness of Bangkok metropolitan public transport by using modified quantitative gap analysis. *International Journal of Built Environment and Sustainability, 7*(1), 81–97.

Peungnumsai, A., Miyazaki, H., Witayangkurn, A., & Kim, S. M. (2020). A Grid-based spatial analysis for detecting supply-demand gaps of public transports: A case study of the Bangkok metropolitan region. *Sustainability, 12*(24), 10382.

Potter, K., & Olaoye, F. (2024). Citizen engagement and participatory governance in smart cities. In *EasyChair.*

Sao, A., & Gupta, J. (2023). Sustainability indicators and ten smart cities review. In *2023 IEEE International Conference on Contemporary Computing and Communications (InC4)* (Vol. 1, pp. 1–6). IEEE.

Siangsuebchart, S., Ninsawat, S., Witayangkurn, A., Pravinvongvuth, S. (2021). Public transport GPS probe and rail gate data for assessing the pattern of human mobility in the Bangkok metropolitan region, Thailand. *Sustainability, 13*(4), 2178.

Thaithatkul, P., Sanghatawatana, P., Anuchitchanchai, O., Laosinwattana, W., Liang, J., & Chalermpong, S. (2023). Travel behavior change of public transport users during the COVID-19 pandemic: Evidence from Bangkok. *Asian Transport Studies, 9*, Article 100102.

Vempati, S., & Nalini, N. (2024). Securing Smart Cities: A cybersecurity perspective on integrating IoT, AI, and machine learning for digital twin creation. *Journal of Electrical Systems, 20*(3), 1420–1429.

Wajid, M. S., Terashima-Marin, H., Najafirad, P., Pablos, S. E. C., & Wajid, M. A. (2024). DTwin-TEC: An AI-based TEC district digital twin and emulating security events by leveraging knowledge graph. *Journal of Open Innovation: Technology, Market, and Complexity, 10*(2), Article 100297.

Wimolwattanapun, W., Hopke, P. K., & Pongkiatkul, P. (2011). Source apportionment and potential source locations of PM2.5 and PM2.5–10 at residential sites in metropolitan Bangkok. *Atmospheric Pollution Research, 2*(2), 172–181.

World Population Review. (2024). *Bangkok population*. Retrieved July 15, from https://worldpopulationreview.com/world-cities/bangkok-population

Xiao, M., Chen, L., Feng, H., Peng, Z., & Long, Q. (2024). Sustainable and robust route planning scheme for smart city public transport based on multi-objective optimization: Digital twin model. *Sustainable Energy Technologies and Assessments, 65*, Article 103787.

Mahdi Aghaabbasi is a postdoctoral research fellow at the University of Central Florida (UCF), where he's part of the Urban Digital Twin Lab team. His research covers a mix of transportation planning, machine learning, transport policy, and urban & regional planning. He has published quite a number of research papers and has been cited over 775 times—a sign that his work's making waves in the academic world. Mahdi's also active in the research community, often sharing insights at events, including a recent talk at the IRF Global R2T Conference and Exhibition in December 2024.

Soheil Sabri is an urban planner and geospatial scientist who's passionate about improving cities through research, hands-on work, and teaching. Currently, he is leading the Urban Digital Twin Lab and also teaching as an Assistant Professor at the School of Modelling, Simulation, and Training (SMST) at the University of Central Florida in the U.S. His work mainly revolves around Urban Digital Twins—basically smart, digital versions of cities—along with 3D/4D planning tools, geo-simulation, and data analysis. All of it's aimed at helping urban planners and decision-makers make smarter, more informed choices for cities of the future.

Chapter 6
Smart City Adoption and the Politics of Urban Futures: A Case Study of Magelang City, Indonesia

Clusters of Future Studies: Past and Futures; Environmental Futures; Technological Trends

Arif Budy Pratama

Abstract Previous studies have reported that local governments play an essential role and are considered the leading actors in executing smart city initiatives. However, the motives and decision-making rationale behind smart city planning and implementation have yet to be clearly understood. This chapter empirically explores the motivation behind the adoption of smart cities. A case study focusing on the Magelang Smart City development initiative in Indonesia from 2013 to 2021 was employed to address the research objective. The findings showed that the smart city adoption was not purely motivated by public goods but also embedded within political and bureaucratic benefits. Drawing from the findings, a dual motive framework for the smart city adoption was proposed. It can also be regarded as a heuristic approach to understanding why a city adopts a smart city concept in its urban development strategy. The framework entails push and pull factors that drive a city's pursuit of being smart. This study contributes to the debates of smart city scholarships, highlighting the political and social dimensions in the Asian context.

Keywords Smart cities · Urban development · Urban governance · Indonesia · Asia · Future cities

6.1 Introduction

The latest report from UN-Habitat (2018) reveals that despite progress in addressing the 2030 Agenda for Sustainable Development (SDG), many challenges still need to be addressed in achieving the urban development targets. Incorporating governance

A. B. Pratama (✉)
Center for Smart and Sustainable Cities, Universitas Tidar, Magelang, Indonesia
e-mail: arifpratama@untidar.ac.id

R. Zainol et al. (eds.), *Future of Cities in Asia*,
https://doi.org/10.1007/978-3-032-01016-2_6

capacity and (information) technological innovation, smart cities hold new hopes for tackling various urban problems as a consequence of the rapidly urbanising world. Regarding global urban governance and development, smart cities address SDG 11 targets and indicators (Klopp & Petretta, 2017; Wendling et al., 2018).

The potential of smart cities has attracted politicians, policymakers, urban planners, and industries in the Global North and Global South alike (Joo & Tan, 2020a, 2020b). Smart city projects have mushroomed in Asia as a new trend in urban planning. The Indonesian 100 Smart Cities movement, which has been implemented since 2017, indicates the Indonesian government's commitment to developing smart cities throughout the country. Indeed, several cities in Indonesia initiated the smart city program before the central government inaugurated the 100 Smart Cities movement (Pratama, 2018; Supangkat, 2015). Until 2020, the national program has facilitated 99 Indonesian cities[1] (Ditjen Aplikasi Informatika, 2022). Magelang City has been appointed by the Ministry of Communication and Informatics as one of the pilot projects for the 100 Smart Cities movement. Figure 6.1 shows the map of Magelang City, located in the Central Java Province. The city encompasses three subdistricts and 17 urban villages covering 18.32 km^2. Within a population of 121,872 people, the city boasts a population density of approximately 6633 people per square kilometre (BPS Kota Magelang, 2021). Despite its relatively small size, Magelang City has gained recognition for its exemplary governance practices. For example, the human development index is 80.39, one of Indonesia's highest (BPS Kota Magelang, 2023). In the area of regional competitiveness, this city also received recognition for its success in building its competitive ecosystem. The Central Java Provincial Government recognised Magelang City as one of the best competitive cities in the province (Pemerintah Provinsi Jawa Tengah, 2019b). The regional competitive advantage has also contributed to the effort to reduce poverty. According to data from the National Socio-economic Survey, the city was the top performer in the Central Java Province in poverty alleviation (BPS Kota Magelang, 2021).

Magelang City has been recognised for its smart city initiatives. It won the Indonesian Smart City Rating (RKCI) four consecutive times in 2015, 2017, 2019, and 2021 for the category of small cities. In the assessment, a small city is defined by the maximum population living on the city's territory, which is less than 200,000 people. Several smart city facilities have been implemented in the city of Magelang, including a command-control room and Area Traffic Control System (ATCS) for managing the traffic, DATAGO for government data management system, a Magelang Smart City (Magesty) for smart mobile app, and various other digitised public services. For example, ATCS has improved road users' compliance regarding to traffic regulations and economic efficiency (Prasetyo, 2017). The ATCS technology is integrated with the traffic management system as part of the smart mobility dimension.

As many cities in Indonesia have claimed to be smart cities, the motivations driving these initiatives remain understudied. This chapter aims to explore the motivation behind the smart city adoption policy. It seeks to answer the following question:

[1] Due to the COVID-19 pandemic, several national and sub-national government programs and budgets were revised and refocused to address the pandemic issues.

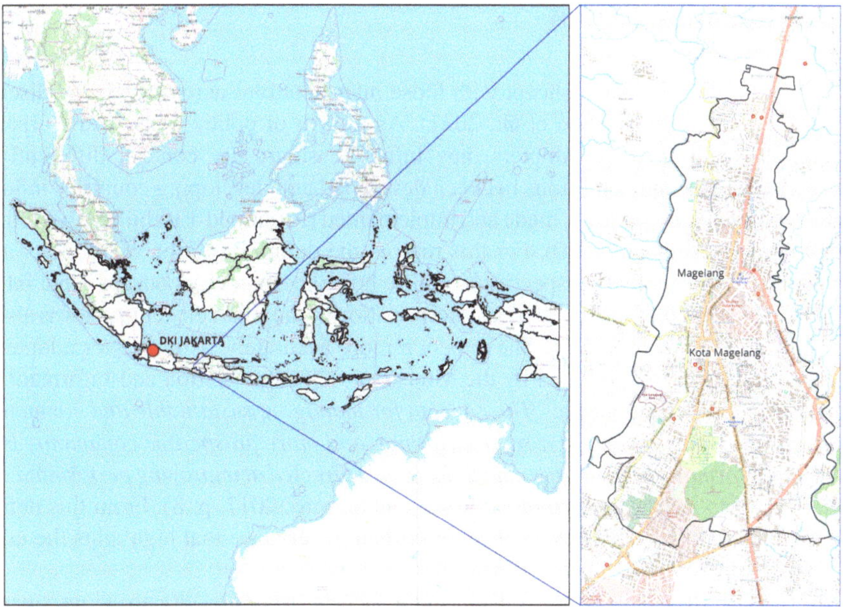

Fig. 6.1 The map of Magelang City, Indonesia. *Source* OpenStreetMap (https://tile.openstreetmap.org/) and was modified with QGIS, March 30, 2024.

Why does Magelang City adopt the smart city concept as its urban development strategy? This crucial question, yet understudied, is essential to unravel the impetus for adoption within the urban political-administration spheres. Focusing the investigation on one city is beneficial since it allows to explore the study object within its real-life context based on the empirical investigation. Answering this puzzle provides nuanced ideas to advance our knowledge of smart city adoption patterns in the Asian context. This study provides important insights into the adoption processes that shape the formulation of smart city policies, program planning, and project execution. It also unpacks the socio-political dynamics inside the city government institution. This chapter proposes a dual motive adoption model that encompasses push and pull factors motivating city governments adopt smart city concept in urban development. It contributes to the debate on smart city scholarships, especially the politics and policy of smart city adoption in the Global South.

This chapter is organised as follow: The next section presents literature review followed by the methodological approach employed during the fieldwork in Magelang City as research locus. It then moves to the main analysis, explicating and discussing the adoption motives before ending with the conclusion.

6.2 Literature Review

The smart city is a vague concept with loose interpretations across academic disciplines (Pratama, 2021b; Soe et al., 2021). As a form of policy transfer and urban strategy travelling ideas, adopters may infer the smart city concept differently. Diverse contexts and conditions between developed and developing countries influence how smart city policy is made and implemented (Fromhold-Eisebith & Eisebith, 2019; Tan & Taeihagh, 2020). For this reason, the definition of the smart city was framed from the applied perspective to realise how the smart city concept is put into practice. In addition, a practical local interpretation of the smart city concept was also employed to contextualise the phenomenon more accurately. This chapter adopted the official definition provided by the Ministry of Communication and Informatics of the Republic of Indonesia: "*The concept for managing sustainable and competitive cities/regions to achieve a safer, healthier, and more prosperous community by utilizing information technology and innovations directed at improving performance, efficiency, and citizen participation.*" (Kemenkominfo, 2017, p. 8). From this definition, the smart city denotes a concept in urban governance that highlights the use of technology to achieve better urban life.

Several studies have identified the main factors driving a city to adopt smart urbanism. For instance, Yigitcanlar et al. (2018) proposed a conceptual framework that links the main drivers to smart cities' outcomes based on a systematic review. Their framework consists of community, technology, and policy. In a more detailed analysis, Tan and Taeihagh (2020, p. 11) synthesized eight driving forces for smart city adoption in developing countries, including government's financial capacity, sound regulation, information technology and infrastructure readiness, economic stability, human capital, citizen engagement, knowledge transfer and participation from non-government sectors, and supportive environment that promotes innovation. Both reviewed the normative-formal drivers, thus failing to unpack the underlying latent motives and address political dynamics inside the city hall.

From an empirical analysis, Dowling et al. (2021) argued that making smart cities is more or less a craft rather than fully rational policy-making. Their multiple case studies in the Sydney Metropolitan Region, Australia, have shown the interplay of a priori drivers, ad hoc initiatives, and post hoc strategies during smart city development. It suggests smart city policy-making requires a deliberative strategy to respond to non-linear processes and recursive drivers embedded in political-economy interests (Pratama, 2023). As such, there is a call for an in-depth investigation to uncover the latent interests behind smart city initiatives.

Although many Asian cities have declared their smartness, the rationale for decision-making behind smart city planning and implementation has yet to be clearly understood. More attention should be paid to such a fundamental discussion focusing on state actor since previous studies on Asian smart cities have reported that both national and local governments play an essential role and are considered the leading actors in executing smart city initiatives (Han, 2023; Joo & Tan, 2020a, 2020b; Joo, 2023; Pratama, 2018; Pratama & Imawan, 2019, 2020). Bringing the case of a local government in an Indonesian city allows a thorough and comprehensive empirical investigation on political dynamics in urban governance drawing from policy actors' real-life experiences.

6.3 Methodology

This empirical study employed a case study research design, focusing on Magelang's smart city development from 2013 to 2021. The case study approach allows us to comprehensively understand the studied phenomena by providing an in-depth and contextual analysis. Fieldwork was conducted to collect primary and secondary data from August 2020 to January 2021. The primary data was drawn from two expert interviews, six semi-structured in-depth interviews, and two speeches from the Magelang City mayor. The interviewees for the semi-structured interviews comprised the City Legislative Council members, the Magelang City Government's high and middle-level civil servants, and the information, communication, and technology companies involved in the Magelang smart city projects. Interviewees agreed that their real names should be confidential and anonymous. Table 6.1 (see the next page) provides information about the data collection, including data collection methods, interviewees and their descriptions, and data collection dates.

To complement the primary data, five policy documents related to the Magelang Smart City initiative as the secondary data sources were analysed. The policy documents included the following: Magelang City Medium-Term Development Planning 2016–2021, Magelang Smart City Masterplan 2016–2021, Magelang Smart City Masterplan 2019–2025, Smart City Master Plan Guide from the Ministry of Communication and Informatics 2017, Mayor of Magelang City Decree on Smart City Council 2019.

6.4 Findings

The translation from idea to policy requires a set of rationales in which the Magelang city government pushes the concept of smart urbanism forward into policymaking. The politics of policy-making provides a politico-legal foundation through which the city government further translates and executes the strategic policy into managerial and operational layers called programs (Indonesian: *program pembangunan*) and projects (Indonesian: *kegiatan*). The analysis has shown that the drivers of smart city adoption were not purely motivated by public goods; however, it is also embedded within political and bureaucratic benefits.

The reason behind smart city adoption can be viewed as dual motives, including push-and-pull factors. The push factor refers to the city conditions and contextual attributes that encourage the city government to integrate smart city principles to govern the city's urban issues. This factor is context-based, unique, and typical,

Table 6.1 Data collection methods, interviewees and their descriptions, and data collection dates

Data collection methods	Interviewees	Description of interviewees	Data collection date
Expert interview	Expert 1 from Institut Teknologi Bandung	Magelang Smart City's project advisor since 2015–2019	4 September 2020
	Expert 2 from Universitas Muhammadiyah Magelang	An academician who studies smart waste systems in Magelang City	14 September 2020
Semi-structured interview	Interviewee 1	A civil servant from Magelang City Government Research and Development Agency	28 September 2020
	Interviewee 2	A civil servant from the Office of Communication, Informatics, and Statistics	5 October 2020
	Interviewee 3	A civil servant from the City Development Planning Agency	5 October 2020
	Interviewees 4 and 5	Two members of the City Legislative Council as political actors in the Magelang Smart City development projects	6 and 11 November 2020
	Interviewee 6	A consultant from Gama Techno, a smart city enterprise. A smart city enterprise hired by the Magelang city government	25 January 2021
Public speeches*	Magelang City Mayor*	A political actor and top leader of the Magelang Smart City Programme	18 and 26 October 2020

* The last one represents a different mode of data collection, i.e. recorded public speeches, which is not interviews like the rest. It is included in the table as it still serves as a data source, albeit the only exception, while the majority of the data was collected through interviews with interviewees

with each city's characteristics. In comparison, the pull factor implies the expected outcomes that attract city governments to smart city adoption. It deals with the future agenda of urban development in Magelang City.

6.4.1 Push Factors of the Smart City Adoption

Four contextual attributes as the push factors can be identified. Firstly, Magelang City grapples with common urban problems, including low quality of public services, traffic congestion, inadequate sanitation and housing infrastructure, waste issues, and inefficient energy consumption.[2] For instance, the latest data from the Magelang City Office of Environmental Services (2022), shows the increasing volume of household waste and the declining capacity of the waste collection process. In addition, 24.7% of households have their drinking water source located less than 10 m from their faeces storage (BPS Kota Magelang, 2023). This data indicates that suboptimal sanitation conditions were apparent in Magelang City.

Magelang City has limited natural resources, which served as the second motivation. According to the latest data from the Magelang Statistics Agency, there were minimal exploration and exploitation activities in the mining sectors in the city (BPS Kota Magelang, 2021). It was indicated by the absence of companies, zero numbers of workers, and no records of investments in the mining and energy sectors. Unlike surrounding areas reliant on agricultural activities and eco-tourisms, Magelang City serves as the hinterland's hub, prompting the Magelang City government to emphasise its role as a service provider and promote a service city' as its tagline.[3] The scarcity of natural resources has compelled the city government to prioritise delivering high-quality public services to both residents of Magelang City and neighbouring cities.

The third contributing factor was the strategic location of Magelang City. The strategic location encourages the adoption of the smart city concept to capitalise on its potential for commercialisation and stimulate economic activities. Despite its modest size and geographical features, its position on the border of Yogyakarta and the Central Java Province makes it a crucial hub in transportation, trade, economic, creative industries, and tourism activities between the two provinces. This strategic significance is underscored by Presidential Regulation No. 58 Year 2014 and No. 3 Year 2016 as a buffer for the Strategic National Project for Borobudur Temple as one of the UNESCO heritage sites and international tourism destinations.

Furthermore, the strategic location is also linked with the fourth contributing factor, whereby the Central Java Provincial Government gives a mandate to Magelang City as the Regional Centre Activity. This designation empowers the city to serve as the regional centre for economic development, including export–import, manufacturing, transportation, and other industries. In practice, Magelang city supported and contributed significantly to the regional economy of surrounding regencies known as Purwomanggung, the abbreviation for *Purworejo, Wonosobo, Temanggung, and the Magelang regency* (Pemerintah Provinsi Jawa Tengah, 2019a). In fulfilling the role of regional activity, the Magelang City government provides diverse public services

[2] This factor has been extensively stated in the Magelang Smart City Masterplan 2016–2021, Magelang Smart City Masterplan 2019–2025(Bappeda Kota Magelang, 2016a; Diskominsta, 2019) and also supported by two expert interviews.

[3] Most interviewees stated the role of Magelang City as the service provider in the region.

spanning from education, financing, industries, transportation, health, trade, and administrative service domains. The Magelang City government has also enhanced its digital infrastructure and capacity to bolster the Regional Centre Activity (Bappeda Kota Magelang, 2016b). In this sense, the positioning of Magelang Smart City leverages local economic development in the regional areas of Yogyakarta and Central Java Province. The case of Magelang Smart City exposes the developmentalism of smart cities as observed in Seoul and Singapore at the Global Cities context (Joo, 2023).

6.4.2 *Pull Factors of the Smart City Adoption*

Apart from the push factors, the case study identified two pivotal aspects enticing the city government to embrace the smart city initiatives. The first encompasses a normative factor aligning with the formal objectives of policies in smart city development. Ideally, the smart city program aims to solve urban problems innovatively. This objective signifies the dimensions and indicators in Magelang City's development planning document. One of the most emerging normative objectives is modernising the city government's bureaucracy in delivering urban service.

In regards to the smart city programme, the city government has implemented various e-government applications such as the Magelang Smart City mobile application, civil registration system, business licensing system, one-data platform, and local government financial systems such as e-planning, e-budgeting, and e-monitoring. From the City Council's perspective, the bureaucratic process's modernisation is urgently required. "*Yes, we are now in the globalization era, do not be left behind. We are here to catch the fourth industrial revolution, the 4.0 era, in which bureaucracy has to adapt.*" (Interview with a Magelang City Legislative Council member, 11 November 2020).

Modernising the public service enables government bureaucracy to improve public service provision. The administration introduced the excellence services credo within the Magelang's smart government dimension by developing One-Stop Services (OSS) in most city offices. One example of OSS implementation was the investment and business license. Simplifying the process of investment and business licensing not only enhances the city's competitiveness but also stimulates economic development in the city areas. With more businesses established in Magelang City, economic activities will flourish, boosting the city's revenue from local taxes. In this sense, the adoption motives followed techno-centric smart city concept implemented in other Asian cities such as China, India, and South Korea (Han, 2023).

The next normative objective was to achieve a liveable city for its residents.[4] The smart city policymaking process prioritised the themes: innovative, digital, green, and sustainability. The city aims to govern city infrastructures effectively by utilising

[4] Most interviewees and policy documents gave similar insights into the normative motivation to achieve a more livable city.

various sensor-based technologies. For example, the city set the ideal conditions for an information system to conduct a monitoring system to maintain 30% of the public green space. Additionally, some efforts have been executed to decrease water and air pollution, such as a control room that monitors the pollution issue.[5] Further projects supporting the realisation of liveable cities include Geographic Information System (GIS) based technology to support building licenses and integrating alternative energy generators throughout Magelang City (Bappeda Kota Magelang, 2016b).

In addition to normative motives, the Magelang Smart City development has also been motivated by latent interests. The study revealed four latent motivations driving smart city adoption in the research locus. The first leading motive was to fulfil the Magelang City Mayor's political promise. Adopting the smart city concept was one of the key components of his political campaigns in 2016's mayoral election. An interview with a participant from the City Development Planning Agency revealed that the current mayor deliberated with the agency regarding his vision for his second term during the 2016 mayoral candidacy and campaign.

> Before the 2016 mayoral election, the mayor came to me and asked for suggestions for his vision and mission for the election. I proposed the smart city theme as one of the visions, and thanks to God, he won, and the vision of Magelang City 2016–2021 was Magelang city as a modern and smart service city based on a prosperous and religious society. (Interview with the Head of the City Development Planning Agency, 5 October 2020).

Consequently, Magelang City Medium-Term Development Planning 2016–2021 incorporates the smart city approach as its urban development priority. The smart city theme was explicitly articulated as the mayor's vision. This guides the strategies in urban development throughout the mayor's five-year administration. Implementing smart city projects signifies that the elected mayor has fulfilled his political commitment. Adopting the smart city was inherently political since it represented the incumbent mayor's political agenda during his second term. The emphasis on digitalisation, modernisation of civil services, and the service city within the frame of smart city development all enhance the incumbent mayor's political legitimacy.

Secondly, the interest in adopting the smart city concept stemmed from the desire for recognition and prestige. Projects in smart cities often result in the development of city infrastructures that can enhance personal recognition,[6] especially during the mayor's tenure. The mayor's speeches on 26 October 2020 declared the intention to achieve the city's recognition as the 'Singapore of Indonesia'. "*It does not matter to be the smallest city, but we are on the way to as successful as Singapore. We are Singapore in Central Java*" (Mayor speech, 26 October 2020). His statement was also amplified by a Magelang City Legislative Council member who said that "*…small is beautiful. That is why we will develop our city as Singapore did*" (Interview, 11 November 2020).

---

[5] Interview with civil servants from the Development Planning Agency and Office of Communication, Informatics, and Statistics.

[6] Some interviews with experts and civil servants revealed the side effects of smart city infrastructure and mitigated the smart city's impact for political benefits.

Regarding personal branding, the mayor has been awarded the National Smart City Award for four consecutive years: 2015, 2017, 2019, and 2021. These swayed the mayor's reputation and boosted political marketing since many outlets broadcast the awards enormously. One of the most obvious was the city billboard showing the mayor receiving the awards in strategic places around the city. In addition, many city infrastructures, such as the traffic management system, command centre, Magelang DATAGO, and Magelang Cerdas apps, were some of the legacy worth remembering as the footprints of the mayor's performance in digitalisation projects.

Third, the adoption motive was closely related to the city branding strategy. Given the intensified urban competition, many cities, including Magelang City, have put smart cities as one of the tools in city branding strategies. Based on marketing studies (Kavaratzis, 2018), city branding aims to expose a positive image (Govers, 2012) and promote local competitive advantages (Boyle, 1997). The recognitions and smart city awards were aimed to boost the city's image as a modern, safe, and liveable city. From a more pragmatic point of view, the smart city as city branding was an entry point that attracted business investment and tourism. An interview with the participant from the Office of Communication, Informatics, and Statistics unveils that the smart government and smart economy supported by digital technologies would build a good city image and attract investment. He stated, "*...the recognition of smart cities enhanced citizens' trust and made Magelang an investor-friendly city. This city has efficient bureaucracies and does not hinder investment*" (Interview, 5 October 2020).

Lastly, the adoption of smart city policy also provided benefits to the city government bureaucracies. These benefits align with the budget-maximising model, where city government bureaucracies seek to maximise their budget, resources, and autonomies for their interests (Niskanen, 1994). These interests included personal utilities, power, and status. During an Interview with the Magelang City Legislative Council member on 26 October 2020, it was evident that the proposal for smart city development was linked with the politics of city budgeting. In practice, the city government units often compete for budget shares, personnel allocation, and the authority to manage and execute local development projects. The interviewees stated "*...every executive development program is political as it represents how city government accommodate interest through budget...*" (Magelang City Legislative Council member, 6 November 2020). In the context of smart city projects, city bureaucracies can propose and execute many projects related to smart city initiatives. They carried out these projects to realise budget allocation, aiming to contribute to economic growth and enhance government expenditure using the public budget. Figure 6.2 portrays the dual motive of smart city adoption in the case of Magelang Smart City initiatives.

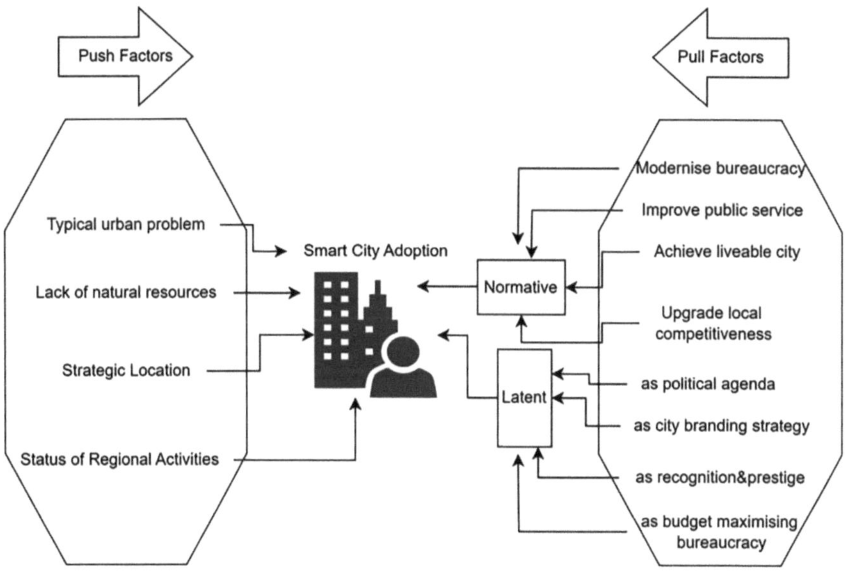

Fig. 6.2 The dual motive framework (push and pull factors) for smart city adoption

6.5 Discussion of Implications

The findings offer a more comprehensive and nuanced analysis, revealing the multi-faceted attributes of smart city adoption drivers that have previously been overlooked. By shedding light on latent factors, this study enhances our understanding of the dynamics and complexities of smart city adoption motives. It serves as a valuable complement to previous research that mainly focused on normative factors (*inter alia* Tan & Taeihagh, 2020; Yigitcanlar et al., 2018). While normative drivers tend to capture the visible aspects of urban politics and local governance, latent motivations uncover the underlying dynamics of decision-making processes during the policy design and implementation. In the case study, the latent political motives behind smart city development highlight the significant influence of mayor's political authority in shaping decision-making process.

Additionally, the case study underscores the non-linear nature of smart city policy-making, highlighting the political dynamics and deliberative strategies during the adoption processes (Dowling et al., 2021; Pratama, 2023). In that sense, smart city development should not be regarded as technocratic projects per se, instead, local authorities must consider the socio-political facets of smart city planning. For instance, the Magelang City Mayor's political promise led to political decisions and resource allocations to Magelang Smart City program in the Magelang City Medium-Term Development Planning 2016–2021. The adoption of smart concept, in this case, was heavily driven by political agendas, often creating political pressure on the bureaucracy. Empirical evidence suggests that such political pressure may contribute to premature policy implementation due to inadequate bureaucratic capacity and technological infrastructure (Pratama, 2024).

Furthermore, the motivation to use smart city recognition as a city branding strategy can be viewed as a double-edged sword. On one side, it creates urban identity by positioning the city as technologically advanced, sustainable and liveable (Grebosz-Krawczyk, 2021; Wang, 2023). On the other hand, the motive for recognition as part of city branding strategies carries the risk of superficial urban development. Driven by smart city awards, many cities invested in digital technologies with limited tangible outcomes for urban development (Pratama, 2021a).

6.6 Future Directions

Considering the implications of latent motives, future smart city planning should strike a balance and maintain the tension between push and pull factors in smart city adoption. As Dowling et al. (2021) emphasise on political dynamics, political legitimacy from the political actors matters during agenda setting and policy formulation. Subsequently, urban planners must also account for multi-faceted objectives and interests of various stakeholders when designing smart city master plans. In addition, city governments are required to synchronise the socio-political, administrative, economy, and technological aspects in their smart city development plans.

In the Asian context, the smart city concept has been an urban policy fad that shaped future trajectories of urban planning (Phan & Damian, 2022; Biswas & Dygas, 2023; Joo & Tan, 2020a, 2020b). The case report has enriched the variations of smart city adoption, highlighting the political economy dynamics in smart city policymaking. The future aspiration of smart city development, thus, should consider the humanity dimensions (Hitachi-UTokyo Laboratory, 2020), environmentally-adaptive (Han, 2023), and changing pattern of public demands (Kim, 2022). These should be incorporated within the smart cities' design facilitated by technological infrastructures.

6.7 Conclusion

The analysis provided insights into the impetus for smart city adoption in one of the Asian cities. The adoption in the case study was not purely motivated by public goods but also embedded in interests situated within political and bureaucratic benefits. The adoption driver elucidates the rationale behind the motive, encompassing both push and pull factors.

Push factors denote the city circumstances and contextual attributes that prompt the city government to integrate smart city principles in addressing urban issues. These factors are usually context-based, unique, and typical. In the case study, push factors relate to 'classic' urban problems, inadequate urban resources, the city's location, and the status of the regional economic centre. Pull factors imply the expected outcomes that attract city governments to adopt the smart city concept. These represent the future projections and city imaginations that lead to the realisation of liveable cities and communities. In the case study, pull factors can be normative and latent. Expectations such as modernising the city's bureaucracy, improving the quality of public services, and upgrading local competitiveness have been identified as the normative motives. Moreover, smart city adoption harboured latent motives that aim to benefit political and bureaucratic actors. These include manoeuvres for incumbent political currency, city branding, and a means of budget maximising for bureaucracy.

Based on the exploratory findings, the study proposes a dual motive framework for the smart city adoption model. It offers a heuristic lens to understand a city's rationale for adopting a smart city initiative. This framework encapsulates push and pull factors influencing a city's pursuit of smartness and aspiration of the "smart" status and recognition. Lastly, the chapter notes that the framework has limitations since it draws from a single case study, which hinders the generalisation of findings. Future endeavours may expand the scope by conducting multiple comparative studies across diverse cities and countries.

Highlights

- Address diverse stakeholders' interests in Asian urban planning to navigate power relations and ensure inclusiveness.
- Avoid technocratic bias in future Asian smart cities to accommodate context-driven planning and promote stakeholders' engagement.
- Balance push–pull factors to reconcile varied motives in smart city adoption.

References

Bappeda Kota Magelang [Regional Development Planning Agency of Magelang City]. (2016a). *Laporan Akhir Magelang smart city Penyusunan Dokumen masterplan smart city Kota Magelang [Final Report of Magelang Smart City Masterplan Document)].* Kota Magelang.

Bappeda Kota Magelang [Regional Development Planning Agency of Magelang City]. (2016b). *Master plan Magelang smart city [Magelang Smart City Masterplan].* Kota Magelang.

Biswas, P. K., & Dygas, R. (2023). *Smart cities in Europe and Asia: Urban planning and management for a sustainable future.* Taylor & Francis.

Boyle, M. (1997). Civic boosterism in the politics of local economic development—Institutional positions; and strategic orientations in the consumption of hallmark events. *Environment and Planning a: Economy and Space, 29*(11), 1975–1997. https://doi.org/10.1068/a291975

BPS Kota Magelang [Magelang City Statistics]. (2021). *Kota Magelang dalam angka [Magelang City in Numbers]*. Kota Magelang. http://magelangkota.bps.go.id/website/pdf_publikasi/Kota-Magelang-Dalam-Angka-2015.pdf

BPS Kota Magelang [Magelang City Statistics]. (2023). *Statistik Daerah Kota Magelang 2023 [Statistics Magelang City 2023]*. Kota Magelang.

Diskominsta [Office of Communication, Informatics, and Statistics]. (2019). *Master plan Magelang smart city*. Kota Magelang.

Ditjen Aplikasi Informatika [Directorate General for Informatics Application]. (2022). *Gerakan Menuju 100 smart city [100 Smart Cities Movement]*. https://aptika.kominfo.go.id/2022/07/gerakan-menuju-100-smart-city-2/

Dowling, R., McGuirk, P., Maalsen, S., & Sadowski, J. (2021). How smart cities are made: A priori, ad hoc and post hoc drivers of smart city implementation in Sydney, Australia. *Urban Studies, 58*(16), 3299–3315. https://doi.org/10.1177/0042098020986292

Fromhold-Eisebith, M., & Eisebith, G. (2019). What can smart city policies in emerging economies actually achieve? Conceptual considerations and empirical insights from India. *World Development, 123*. https://doi.org/10.1016/j.worlddev.2019.104614

Govers, R. (2012). Brand Dubai and its competitors in the Middle East: An image and reputation analysis. *Place Branding and Public Diplomacy, 8*(1), 48–57. https://doi.org/10.1057/pb.2011.30

Grebosz-Krawczyk, M. (2021). Place branding (r)evolution: The management of the smart city's brand. *Place Branding and Public Diplomacy, 17*(1), 93–104. https://doi.org/10.1057/s41254-020-00167-2

Han, H. (2023). Smart cities in Asia: Ambiguity, innovation, and evolution. In *Routledge handbook of Asian cities* (pp. 26–37). https://doi.org/10.4324/9781003256533-3

Hitachi-UTokyo Laboratory. (2020). *Society 5.0: A people-centric super-smart society*. Springer Open. https://library.oapen.org/bitstream/handle/20.500.12657/41719/2020_Book_Society50.pdf?sequenc#page=59

Joo, Y. M., & Tan, T. B. (2020a). Smart cities in Asia: An introduction. In Y. Joo & T. Tan (Eds.), *Smart cities in Asia governing development in the era of hyper-connectivity* (pp. 1–17). Edward Elgar Publishing Limited.

Joo, Y. M., & Tan, T. B. (2020b). *Smart cities in Asia: Governing development in the era of hyper-connectivity*. Edward Elgar Publishing.

Joo, Y. M. (2023). Developmentalist smart cities? The cases of Singapore and Seoul. *International Journal of Urban Sciences, 27*(S1), 164–182. https://doi.org/10.1080/12265934.2021.1925143

Kavaratzis, M. (2018). Place branding: Are we any wiser? *Cities, 80*(June), 61–63. https://doi.org/10.1016/j.cities.2018.06.003

Kemenkominfo [The Ministry for Communication and Informatics]. (2017). *Buku Panduan Penyusunan Masterplan Smart City 2017 Gerakan Menuju 100 Smart City [Smart City 2017 Masterplan toward 100 Smart Cities Movement]*. Jakarta.

Kim, J. (2022). Smart city trends: A focus on 5 countries and 15 companies. *Cities, 123*, 103551. https://doi.org/10.1016/j.cities.2021.103551

Klopp, J. M., & Petretta, D. L. (2017). The urban sustainable development goal: Indicators, complexity and the politics of measuring cities. *Cities, 63*, 92–97. https://doi.org/10.1016/j.cities.2016.12.019

Magelang City Office of Environmental Services. (2022). *Volume Sampah Rata-rata per Hari (m3) Menurut Wilayah di Kota Magelang [Average Waste Volume per Day (m3) by Region in Magelang City]*. Magelang Datago. https://data.magelangkota.go.id/dataset/volume-sampah-rata-rata-per-hari-m3-menurut-wilayah-di-kota-magelang

Niskanen, W. A. (1994). *Bureaucracy and public economics*. Edward Elgar.

Pemerintah Provinsi Jawa Tengah [Central Java Provincial Government]. (2019a). *Kota Magelang Jadi Pusat Pengembangan "Purwomanggung." [Magelang City Becomes the Development Centre of 'Purwomanggung.']*. Berita Daerah. https://jatengprov.go.id/beritadaerah/kota-magelang-jadi-pusat-pengembangan-purwomanggung/

Pemerintah Provinsi Jawa Tengah [Central Java Provincial Government]. (2019b). *Kota Magelang Juara I Indeks Daya Saing Daerah Tingkat Jateng 2019 [Magelang City wins first place in Central Java Regional Competitiveness Index 2019]*. Portal Berita. https://jatengprov.go.id/beritadaerah/kota-magelang-juara-i-indeks-daya-saing-daerah-tingkat-jateng-2019/

Phan, T., & Damian, D. (2022). *Smart cities in Asia: Regulations, problems, and development*. Springer Nature Singapore.

Prasetyo, A. (2017). Economic efficiency of audio announcer installation on Area Traffic Control System (ATCS) In Magelang City. *Jurnal Penelitian Transportasi Darat, 19*(2), 83–94.

Pratama, A. B. (2018). Smart city narrative in Indonesia: Comparing policy documents in four cities. *Public Administration Issues, 2*(6), 65–83. https://doi.org/10.17323/1999-5431-2018-0-6-65-83

Pratama, A. B. (2021a). Smart is not equal to technology: An interview with Suhono Harso Supangkat on the emergence and development of smart cities in Indonesia. *Advances in South-East Asian Studies, 15*(1), 1–7.

Pratama, A. B. (2021b). Knowledge distribution on smart city research. *Journal of Urban Culture Research, 23*(December), 25–43.

Pratama, A. B. (2023). *The social interface of smart city development: An empirical study in Magelang City, Indonesia* [University of Bonn]. https://hdl.handle.net/20.500.11811/11019

Pratama, A. B. (2024). When trust coexists with mistrust: An ethnographic account of the consequences of digitalization project. *International Journal of Public Administration, Online First, 1–11*. https://doi.org/10.1080/01900692.2024.2368253

Pratama, A. B., & Imawan, S. A. (2019). A scale for measuring perceived bureaucratic readiness for smart cities. *Public Administration and Policy: An Asia Pacific Journal, 22*(1), 25–39. https://doi.org/10.1108/PAP-01-2019-0001

Pratama, A. B., & Imawan, S. A. (2020). Bureaucratic readiness for smart city initiatives: A mini study in Yogyakarta City, Indonesia. In Y. Joo & T. Tan (Eds.), *Smart cities in Asia governing development in the era of hyper-connectivity* (pp. 148–161). Edward Elgar Publishing Limited. https://doi.org/10.4337/9781788972888

Soe, R.-M., Schuch de Azambuja, L., Toiskallio, K., Nieminen, M., & Batty, M. (2021). Institutionalising smart city research and innovation: From fuzzy definitions to real-life experiments. *Urban Research & Practice, 00*(00), 1–43. https://doi.org/10.1080/17535069.2021.1998592

Supangkat, S. H. (2015). Smart city development in Indonesia and Asian-African nations. *IECE, 78*. https://www.ieice.org/eng/activities/ieice_global_plaza/2015/78.html

Tan, S. Y., & Taeihagh, A. (2020). Smart city governance in developing countries: A systematic literature review. *Sustainability (Switzerland), 12*(3), 1–29. https://doi.org/10.3390/su12030899

UN-Habitat. (2018). Tracking progress towards inclusive, safe, resilient and sustainable cities and human settlements: SDG 11 SYNTHESIS REPORT HIGH LEVEL POLITICAL FORUM 2018. In *Tracking progress towards inclusive, safe, resilient and sustainable cities and human settlements*. https://doi.org/10.18356/36ff830e-en

Wang, H.-J. (2023). Smart city branding vision: Multiple stakeholder perspectives. *Innovation: The European Journal of Social Science Research*, 1–25. https://doi.org/10.1080/13511610.2023.2296384

Wendling, L. A., Huovila, A., zu Castell-Rüdenhausen, M., Hukkalainen, M., & Airaksinen, M. (2018). Benchmarking nature-based solution and smart city assessment schemes against the sustainable development goal indicator framework. *Frontiers in Environmental Science, 6*(JUL), 1–18. https://doi.org/10.3389/fenvs.2018.00069

Yigitcanlar, T., Kamruzzaman, M., Buys, L., Ioppolo, G., Sabatini-Marques, J., da Costa, E. M., & Yun, J. H. J. (2018). Understanding 'smart cities': Intertwining development drivers with desired outcomes in a multidimensional framework. *Cities, 81*, 145–160. https://doi.org/10.1016/j.cities.2018.04.003

Arif Budy Pratama is an assistant professor at the Faculty of Social and Political Sciences and lead Research Center for Smart and Sustainable Cities in Universitas Tidar, Magelang, Indonesia. His research focuses on political science and public administration, particularly in areas like smart city development and urban governance. He's got a strong academic background—starting with a degree in Public Administration from Universitas Diponegoro, followed by a master's in Strategic Studies from Universitas Indonesia. He then went on to earn another Master's in Public Administration from the University of Exeter and, later, a Ph.D. from the University of Bonn. He has written extensively on smart cities, urban policy, and how digital government builds its reputation.

Chapter 7
Key Drivers and Headwinds for Future Cities in South Asia, East Asia and Southeast Asia

Clusters of Future Studies: Environmental Futures; Technological Trends

Norzailawati Mohd Noor and Rosilawati Zainol

Abstract Future cities in Asia are at the forefront of urban transformation, combining advanced technologies, sustainable practices, and innovative planning to address the challenges of rapid urbanisation. Asia is experiencing a digital transformation, creating a platform for developing innovative urban areas. Technology and sensors that collect or transfer data, among other technology, are fully integrated in these modern towns. This chapter elaborates on the comprehensive situation of future cities in the context of South Asia, East Asia, and Southeast Asia. Focus has been given to the key drivers and headwinds (forces and challenges) that are faced by the countries. The factors that drive future city development have been counted in terms of physical, social, and economic aspects while also focusing on the community, technology and policies. The outcome of these drivers and headwinds will link productivity, sustainability, accessibility, well-being, liveability, and governance for the expected future cities in Asia. This exploration also shows that the urban futures do not reside solely in technological systems but also involve the flexibility and coherence of other systems and how those systems ensure safety and opportunity for all.

Keywords Future cities · Asia · Technological integration · Governance · Southeast Asia

N. M. Noor (✉)
Urban and Regional Planning Department, Kulliyah of Architecture and Environmental Design, International Islamic University Malaysia, Kuala Lumpur, Malaysia
e-mail: norzailawati@iium.edu.my

R. Zainol
Department of Urban and Regional Planning, Faculty of Built Environment, Universiti Malaya, Kuala Lumpur, Malaysia
e-mail: rosilawatizai@um.edu.my

R. Zainol et al. (eds.), *Future of Cities in Asia*,
https://doi.org/10.1007/978-3-032-01016-2_7

7.1 Introduction

In early 2025, we kept asking what future cities are and what the direction or shape of future cities worldwide, particularly in Asia. Cities in Asia and the Pacific face a complex set of interlinked crises, which carry profound implications for the region's social fabric and development prospects (ESCAP, 2023). Asia is the largest and most populous continent on Earth, known for its vast diversity in geography, cultures, economies, and political systems. The Asia geography covers about 44.58 million square kilometres, making it the largest continent, accounting for 30% of the Earth's total land area (Worldatlas, 2024). The Arctic Ocean bounds it to the north, the Pacific Ocean to the east, and the Indian Ocean to the south. At the same time, it is connected to Europe on the west by the Ural Mountains and the Ural River, often considered part of the Eurasian landmass. There is also the Suez Canal, which marks the boundary between Asia and Africa. The countries and/or territories within each region of Asia are listed in Table 7.1.

As we expected, the current urban world nowadays, which is the future of cities in the last decades, will not all be new cities. Future cities will be an addition to and transform existing cities. The confluence of technology with social and economic trends will create issues and prospects that we have not imagined. Current trends are moving in some cities toward efficiency and autonomy. Buildings and their development are becoming more efficient. Renewable energy is growing fast, and integrated renewable energy generation will become a matter of course (Kiss et al., 2014). The development movement in each Asian country will shape a different approach for future cities. Depending on the issues that need to be tackled, that's how they respond and shape their future cities.

7.2 Methodology

This study takes a deep dive into the key forces shaping future cities in Asia, using a qualitative research approach with a narrative twist. This method is a great fit because it helps uncover the layers of complexity behind urbanisation, technological

Table 7.1 Countries and territories categorised by geographical regions in Asia

Regions	Countries and territories
East Asia	China, Japan, South Korea, North Korea, Mongolia, and Taiwan
Southeast Asia	Indonesia, Thailand, Vietnam, the Philippines, Malaysia, Singapore
South Asia	India, Pakistan, Bangladesh, Nepal, Bhutan, and Sri Lanka
Central Asia	Kazakhstan, Uzbekistan, Turkmenistan, Kyrgyzstan, and Tajikistan
West Asia (Middle East)	Includes Turkey, Saudi Arabia, Iran, UAE, and others
North Asia	It largely consists of Siberia, which is part of Russia

breakthroughs, and shifting social landscapes. The narrative approach allows us to bring in different voices and perspectives, painting a richer, more contextual picture of how cities evolve. By weaving together these diverse experiences, we can better understand what's driving urban space transformation (Höffken & Limmer, 2019).

A qualitative search of relevant literature and online sources is conducted. A comprehensive literature review has thus covered the popular labels that emphasise specific notions of the future city and Asia, noting key features and priority areas highlighted under each label. The study also includes examples of proposed future city projects built from starch or in the planning stage, selected as the cases. Several futuristic city projects are mentioned to establish the difference between understanding and perceiving these concepts. There are numerous visionary proposals, but few are stated here in order to explain these kinds of cities. The cases originate from selected parts of Asia countries. Further compilations and analyses are conducted in order to present the current headwinds and key concepts of future cities in Asia.

7.3 Key Concepts and Issues

The "future cities" concept represents a comprehensive vision for urban development that addresses the growing challenges of urbanisation, climate change, technological transformation, and socio-economic inclusion. This approach, adopted worldwide, emphasises creating sustainable, resilient, and liveable urban environments catering to present and future generations. Many future city theories exist worldwide, comprising recent ideas that can be implemented according to the suitability of each country (Kahachi et al., 2024). Discussion on the key concepts of the future city is not limited to the social science discipline, but also encompasses other disciplines such as environmental sciences, engineering, and health sciences. Table 7.2 (see the next page) presents key concepts of future cities and their distribution of literature coverage from various disciplines.

As Southeast Asian cities navigate the complexities of rapid urbanisation, it is essential to examine global trends in future city development that could offer valuable insights for this country. Around the world, the concept of future cities has gained significant attention as urban areas face unprecedented challenges, including rapid population growth, environmental degradation, and the pressures of technological advancement. The imperative to develop sustainable, smart, and resilient urban environments has also become a central focus for governments and urban planners globally. While the approaches to future cities can vary widely depending on regional contexts, available resources, and strategic goals, these countries should have similarities and generally converge on key themes such as sustainability, technological integration, and social inclusivity.

Technological integration is also the cornerstone of future city approaches, particularly in smart city initiatives. Cities such as Tokyo, Barcelona, and Dubai are at the forefront of leveraging technology to improve urban living standards. These cities are known for their innovative use of the Internet of Things (IoT), big data, and artificial

Table 7.2 Key concepts of future cities and their distribution of literature coverage from various disciplines

Future city key concepts	Descriptions	Literature coverage across various disciplines
Smart City	Smart cities leverage technology and data to enhance urban living, improve efficiency, sustainability, and governance, fostering economic growth and citizen well-being	0.99
Intelligent City	Integrates advanced technologies, data analytics, and innovation to optimise infrastructure, governance, and citizen services, ensuring sustainable urban development	0.98
15-min City	A 15-min city ensures all daily needs are accessible within a 15-min walk or bike ride	0.98
Solar City	It prioritises solar energy use, integrating renewable systems for power generation, sustainability, and environmental impact reduction	0.98
Productive City	It fosters innovation, efficient resource use, economic growth, job opportunities, and improved quality of life	0.97
Experimental City	It tests innovative urban solutions, policies, and technologies, fostering adaptability, sustainability, and inclusive community development	0.96
Sharing City	Promotes shared resources, services, and spaces to enhance sustainability, community collaboration, and efficient urban living	0.95
Digital City	Utilises digital technologies, data, and connectivity to enhance urban services, governance, communication, and citizen engagement	0.94
Resilient City	A city that is adaptable, sustainable, and capable of withstanding and recovering from challenges, shocks, and stresses	0.94
Zero-Waste city	A city minimises waste by promoting reuse, recycling, composting, and sustainable consumption, aiming to eliminate landfills	0.94
Big Cata City	Leverages large-scale data collection and analysis to optimise urban planning, services, and decision-making	0.94
Compact City	Promotes high-density, mixed-use development, reducing urban sprawl while enhancing accessibility and sustainability	0.93
Ubiquitous City	Integrates pervasive technology and iot to provide seamless, real-time services and enhance urban living	0.91
Regenerative City	Restores and enhances natural ecosystems, aiming to create a positive environmental and social impact	0.90
Biophilic City	Integrates nature into urban environments, fostering a connection between people and natural ecosystems	0.90

(continued)

Table 7.2 (continued)

Future city key concepts	Descriptions	Literature coverage across various disciplines
Eco-city	Prioritises sustainable practices, reducing environmental impact through efficient resource use, green spaces, and renewable energy	0.89
Knowledge City	Fosters innovation, education, and research, leveraging knowledge and technology to drive sustainable development	0.87
Liveable City	Ensures high quality of life through safety, accessibility, sustainability, and vibrant social and cultural amenities	0.84
Open City	inclusivity, transparency, and accessibility, fostering citizen participation and equitable opportunities for all	0.84
Low-Carbon City	City minimises emissions through sustainable practices and technologies	0.84
Age-Friendly City	City supports all ages with inclusive, accessible environments	0.83
Zero-Carbon City	City eliminates emissions by using renewable energy and sustainability	0.83
Safe City	Prioritises security, resilience, and community well-being for all	0.82
Sustainable City	City balances economic growth, environmental health, and social equity	0.82
Garden City	Integrates green spaces, sustainable planning, and urban living to promote harmony with nature	0.82
Virtual City	Replicates real-world urban environments digitally, fostering interaction, innovation, and immersive experiences online	0.81
Connected City	Integrates technology and infrastructure to enhance communication, efficiency, and urban living experiences seamlessly	0.78
Green City	Prioritises eco-friendly practices, renewable energy, green spaces, and sustainable urban development	0.76
Inclusive City	City ensures equal opportunities, accessibility, and participation for all residents, regardless of differences	0.76
Circular City	The city minimises waste and maximises resource efficiency by reusing recycling, and regenerating materials	0.76
Sponge City	The city manages water sustainably by absorbing, storing, and reusing rainwater to prevent flooding and drought	0.75
Playful City	Promotes creativity, interaction, and joy through engaging public spaces and playful urban design	0.72
Post Carbon City	The city operates without fossil fuels, relying on renewable energy and sustainable practices for development	0.72

(continued)

Table 7.2 (continued)

Future city key concepts	Descriptions	Literature coverage across various disciplines
Information City	City thrives on data, technology, and connectivity to enhance urban planning and citizen engagement	0.68
Competitive City	Fosters innovation, economic growth, and talent attraction to excel in global markets	0.64
Quantum City	City leverages quantum technology for advanced computing, secure communication, and efficient urban management systems	0.62
Creative City	City nurtures innovation, arts, and culture to drive economic growth and community engagement	0.60
Global City	City influences worldwide economics, politics, and culture through connectivity, diversity, and international integration	0.60
Eco-friendly City	City prioritises environmental conservation, renewable energy, sustainable practices, and reduced ecological footprints	0.58
Entrepreneurial City	City fosters innovation, startups, and business growth through supportive policies, infrastructure, and resources	0.52
Techno City	Integrates advanced technologies into urban systems to enhance efficiency, connectivity, and quality of life	0.5
Learning City	Promotes lifelong education, skills development, and knowledge sharing for inclusive and sustainable growth	0.32
Renewable City	The city relies on sustainable energy sources, reducing environmental impact and supporting a green future	0.29

Source Adapted from Kahachi et al. (2024)

intelligence to enhance city management, reduce congestion, and improve public services (Baig et al., 2017; Perez-Montenegro et al., 2018). For example, Tokyo's smart city initiatives include advanced traffic management systems and smart energy grids, which are pivotal in optimising urban operations and reducing environmental impacts. As Southeast Asia undergoes its digital transformation, these examples highlight the potential for integrating similar technologies to address the region's urban challenges. Whereby the deployment of smart technologies could significantly enhance the efficiency and liveability of cities in Malaysia, Vietnam, and the Philippines, particularly in addressing issues related to infrastructure, transportation, and energy use (Bonafoni et al., 2017; Schrenk et al., 2020; Tsoutsa et al., 2024).

Additionally, the emphasis on social inclusivity and human-centric planning is becoming increasingly important in future city planning worldwide. Cities like Amsterdam and Vienna have been focusing on creating inclusive urban environments that prioritise the well-being of their residents. Amsterdam's "City Doughnut" model, for instance, integrates social and economic objectives with ecological limits, ensuring that urban development meets the needs of all citizens without exceeding environmental boundaries. Vienna's emphasis on social housing and public spaces reflects a commitment to equity and accessibility, ensuring that future cities remain liveable for all residents, regardless of socio-economic status. For Southeast Asian cities, where socio-economic disparities often manifest in urban settings, these global examples underscore the importance of incorporating inclusivity into urban planning. Whereas, ensuring that all residents, including those in underserved communities, have access to essential services and opportunities will be crucial for the region's sustainable urban future.

Future cities in South and East Asia are increasingly focusing on smart technology, sustainability, and innovation. In South Asia, initiatives like Amaravati in India are advancing with projects such as the Quantum Valley, aiming to establish the country's largest quantum computing hub through collaborations with IBM, TCS, and L&T. Dhaka's Detailed Area Plan (DAP) 2022–2035 outlines strategies for sustainable urban development and improved governance (Chisty & Aktar, 2022). Colombo Port City in Sri Lanka is being developed as a significant logistics and financial hub, with substantial investments from China and the U.S.. Islamabad Smart City in Pakistan is part of the country's broader smart cities initiative, focusing on integrating advanced technologies to enhance urban living. According to the Islamabad Master Plan, the city is undergoing a comprehensive review to address challenges such as insufficient public utilities and urban sprawl (Hasan, 2020). In East Asia, Songdo in South Korea, part of the Incheon Free Economic Zone, is a pioneering smart city built on reclaimed land incorporates advanced technologies like IoT, AI, and sustainable infrastructure to enhance urban living. The city features a pneumatic waste collection system, extensive sensor networks, and LEED-certified buildings. Songdo also hosts the Green Climate Fund and several international institutions, emphasising its global significance (Peyrard & Gelézeau, 2020; Bartmański et al., 2023; Ahsan et al., 2024; Potera, 2024). Xiong'an New Area in China is designed as a model for high-quality urban development, featuring smart mobility solutions, ecological restoration projects, and the relocation of major institutions from Beijing (EC, 2022; Do Livramento Gonçalves et al., 2022). Tsukuba, Japan, designated as a super city under Japan's national strategy, serves as a hub for scientific research and innovation. The Tsukuba Smart City Consortium collaborates with universities, research institutions, and government agencies to implement cutting-edge technologies addressing societal challenges. Initiatives include smart grid technology, unified city data platforms, and telemedicine services (Nambu & Ishibashi, 2002; Japan Counseling, 2024; TSCC, 2024). Taipei Smart City in Taiwan has embraced the smart city concept by integrating digital technologies to improve public services and urban sustainability. The city's initiatives include the use of digital ID cards for accessing various services, real-time waste management systems, and the transformation of former landfills into solar power plants. Taipei's efforts have earned it a high ranking in the Smart City Index (Rebecca, 2022). These cities demonstrate how integrated urban planning, technology, and sustainability can create livable and efficient environments, addressing the challenges of rapid surbanisation and environmental concerns in the region.

7.4 Discussion of Implications

7.4.1 Future Cities in South Asia, East Asia and Southeast Asia

Amid a global landscape characterised by economic and geopolitical turmoil, Asia, with its rapidly growing urban populations and diverse socio-economic landscapes, has become a focal point for future city initiatives. As the continent undergoes unprecedented urbanisation, cities across Asia are increasingly adopting innovative approaches to address the complex challenges posed by this growth. Using several examples of case studies from selected parts of Asia, we conclude that a few segments relate to the current trends and scenarios in Asia. The region's unique cultural, economic, and environmental contexts shape the approaches, resulting in various strategies tailored to local needs.

South Asia presents a complex landscape. For instance, India's Smart Cities Mission aims to transform 100 cities through technological integration to address infrastructure gaps, environmental degradation, and rapid population growth. However, South Asian countries often face significant challenges such as financial constraints, institutional weaknesses, and the need to balance innovation with cultural preservation.

In terms of smart city development, East Asian countries such as Japan, South Korea, and China are leading the way, driven by their advanced technological infrastructure and significant government investment. It can be seen in Japan's smart city projects, such as those in Tokyo and Yokohama, that focus on integrating technology with everyday urban life to create more efficient, safe, and sustainable environments. Additionally, South Korea's Songdo International Business District is recognised as a smart city that features advanced connectivity, green infrastructure, and the strategic implementation of IoT and AI for urban management. On the other side, China's rapid urbanisation has led to numerous smart city projects, such as those in Shenzhen and Hangzhou, where big data and AI are utilised to enhance public services, manage traffic, and improve environmental sustainability.

Across Asia, the development of smart cities reflects diverse regional trajectories shaped by varying levels of technological advancement, governance capacity, and urban challenges. East Asia—comprising highly developed nations such as Japan, South Korea, and China—is significantly more advanced in implementing smart cities. These countries have integrated cutting-edge technologies like 5G networks, autonomous transportation, and advanced surveillance systems into urban life, setting a benchmark for the rest of Asia.

In Southeast Asia, countries are increasingly embracing the idea of future metropolises, though they generally lag behind their East Asian counterparts in terms of infrastructure and technological maturity. Singapore stands out as a regional leader through its Smart Nation initiative, which utilises artificial intelligence (AI), data analytics, and the Internet of Things (IoT) to improve public safety, energy efficiency, and urban mobility (Hayward, 2021). Other countries like Malaysia, Thailand, and Indonesia are also pursuing smart city developments, albeit at earlier stages.

Unlike East Asia's technologically driven approach or Southeast Asia's emerging momentum, South Asia's progress is more uneven, shaped by broader socio-economic development needs. These contrasts highlight how the future of urbanisation across Asia is being pursued at different speeds and scales, yet all underscore a shared recognition of the need for smarter, more sustainable cities.

7.4.2 Special Attention to Southeast Asia

This section specifically goes deeper on the Southeast Asia region because this region has seen unprecedented urbanisation in recent decades. Few Southeast Asian cities have been at the forefront of this transformation, driven by economic growth, demographic shifts, and technological advancements. However, these cities face significant challenges, including infrastructural deficits, environmental degradation, and socio-economic disparities. The future of many cities is being threatened by climate change, whether by long-term effects or extreme weather catastrophes. It makes running a city challenging, especially when navigating infrastructure plans, urban planning, housing affordability and inequality. The trajectory of future city development in Southeast Asia is significantly influenced by the distinct socio-economic, political, and environmental conditions unique to each country. Malaysia, Vietnam, and the Philippines, though united by overarching regional trends such as rapid urbanisation, economic growth, and the adoption of technological innovations, face challenges and opportunities that shape their approaches to building cities of the future. As these nations advance, their strategies reflect a convergence of ambitious objectives with the imperative to address intricate local realities.

In Malaysia, the concept of future cities is deeply interwoven with the country's national development plans, particularly through Vision 2020 and the subsequent Vision 2030 frameworks. These strategies aim to propel Malaysia into the ranks of high-income nations, with a strong emphasis on sustainable and inclusive urban development. The Malaysian government has prioritised the development of smart cities as a key component of its urban strategy, focusing on the integration of advanced technologies to enhance the efficiency, safety, and sustainability of its urban environments. Kuala Lumpur, as the nation's capital, has been at the forefront of these efforts, with the Kuala Lumpur and Iskandar Smart City initiatives leading the charge. This initiative aims to transform Kuala Lumpur and Iskandar region into a model future city, utilising the Internet of Things (IoT), big data analytics, and artificial intelligence (AI) to streamline urban management, improve public services, and optimise resource use. The city's approach includes the deployment of smart traffic systems to reduce congestion, smart energy grids to promote energy efficiency, and digital platforms to enhance citizen engagement and governance transparency.

Vietnam's approach to future city development is shaped by its rapid economic growth and urbanisation, which have transformed the country's urban landscape over the past few decades. The Vietnamese government has actively embraced the concept of smart cities as a means of addressing the challenges posed by this rapid urban transformation. Major urban centres such as Hanoi, Ho Chi Minh City, and Da Nang have become focal points for the country's smart city initiatives, each with its own set of strategies and priorities. In Hanoi, the government has launched a comprehensive smart city plan that aims to tackle the city's most pressing urban issues, including traffic congestion, pollution, and infrastructure strain. The plan includes the deployment of smart traffic management systems, the expansion of public transportation networks, and the promotion of green building standards. These efforts are complemented by initiatives to enhance e-governance, improve public safety, and increase the efficiency of public services through digital platforms.

Meanwhile, the Philippines is an acronym for future city growth due to its archipelagic geography, varied socio-economic nature, and rapidly growing urban population. The country's approach to future cities has been shaped by urban challenges, including traffic congestion, inadequate infrastructure, and vulnerability to natural disasters. The government has recognised the need to address these challenges through the development of smart, resilient, and sustainable urban areas. Despite Metro Manila, New Clark City, a flagship project under the Philippines' future city agenda, exemplifies this approach. The project aims to alleviate congestion in Metro Manila, one of the world's most densely populated cities, by establishing a new urban hub that is disaster-resilient, environmentally sustainable, and technologically advanced. The design of New Clark City integrates green spaces, smart technologies, and robust infrastructure, with a focus on creating a liveable and future-proof urban environment. In Metro Manila, for instance, efforts are underway to improve public transportation, enhance disaster resilience, and promote digital governance. Nevertheless, the broader application of smart city concepts across the country encounters substantial obstacles.

In Indonesia, Jakarta Smart City uses e-governance and data analytics to improve public safety and urban mobility, while the planned capital, Nusantara, aims to be a model green city with renewable energy and AI-driven services. Thailand's Phuket Smart City integrates sustainable practices like solar energy and waste management into its tourism-centric strategy, while Bangkok Smart City focuses on flood management, traffic optimisation, and green urban spaces. In Vietnam, Ho Chi Minh City Smart City Initiative and Da Nang Smart City prioritise e-governance, intelligent transport systems, and disaster resilience through data analytics and advanced infrastructure. Meanwhile, the Philippines' New Clark City offers a green, disaster-resilient alternative to Metro Manila, leveraging renewable energy and AI-based public services. These projects reflect the region's commitment to sustainable urban development by integrating technology and innovation, showcasing diverse approaches to creating inclusive, resilient, and environmentally conscious cities.

Singapore leads this trend with its iconic high-rise buildings that seamlessly integrate gardens and green spaces. Public–private partnerships (PPP) are instrumental in driving these initiatives, fostering collaboration between governments and private enterprises to secure funding and technical expertise. Cyberjaya in Malaysia is a prime example, and it was developed as a tech hub through successful PPPs. Additionally, digital governance has emerged as a crucial feature, enhancing transparency and fostering citizen engagement through e-governance platforms, such as the citizen feedback systems implemented in Manila and Jakarta. Finally, disaster resilience is a priority, with infrastructure designed to withstand the frequent natural disasters in the region, such as Bangkok's advanced flood management systems and New Clark City's disaster-resilient urban planning. Together, these features reflect a comprehensive and forward-thinking approach to urban development in Southeast Asia, prioritising technology, sustainability, and resilience to create inclusive and liveable cities for the future. (Henderson & Henderson, 2012).

Future Southeast Asian cities are characterised by innovative, sustainable, and resilient urban planning approaches to address the region's unique challenges. Smart Infrastructure lies at the core of these developments, with the integration of IoT and AI into transport, energy, and public services, exemplified by smart traffic management systems in Jakarta and Bangkok. A strong emphasis on green and sustainable development is evident as cities incorporate renewable energy, green buildings, and urban biodiversity into their planning. Southeast Asia is rapidly adopting smart city initiatives to address urban challenges and improve the quality of life for its citizens through technology, sustainability, and innovative planning. Singapore's Smart Nation Initiative leads the way by integrating IoT, autonomous vehicles, and digital infrastructure. The initiative promotes sustainability features such as real-time traffic management, energy-efficient smart public housing, and urban farms. Malaysia has made strides with Iskandar Malaysia, a low-carbon economic region emphasising green transportation and tech-enabled housing, and Cyberjaya, a hub for digital innovation and tech startups.

To this end, the development of future Southeast Asian cities is driven by rapid urbanisation, economic growth, technological advancements, and environmental sustainability, with urban populations projected to reach 390 million by 2030. Urbanisation, fueled by population growth and migration, is reshaping cities like Kuala Lumpur and Johor Bahru in Malaysia, where natural increases and rural-to-urban migration drive expansion. Vietnam's Đổi Mới reforms have transformed cities like Ho Chi Minh City, Hanoi, and Da Nang into industrial and trade hubs, attracting rural migrants and increasing pressure on housing, transportation, and infrastructure. In the Philippines, the Business Process Outsourcing (BPO) industry and overseas worker remittances have fueled urban growth in cities like Manila, Cebu, and Davao. Manila, one of the world's most densely populated cities, attracts migrants seeking opportunities, while Cebu and Davao benefit from a young, educated workforce driving innovation. However, these cities also face challenges, including the need for accessible housing, robust infrastructure, and adequate urban services. Additionally, with a youthful demographic transitioning toward an ageing population, cities like Manila and Cebu are preparing for future healthcare needs, accessible housing, and social services to ensure sustainability. As Southeast Asia's urban centres evolve, balancing the demands of rapid urbanisation with infrastructure, environmental, and demographic considerations will be key to shaping resilient and inclusive future cities.

Economic growth drives urban development in Southeast Asia, enabling substantial investments in transportation, energy, and digital infrastructure. Countries like Indonesia, Vietnam, and the Philippines are leveraging rising incomes to modernise cities, enhance connectivity, and create employment opportunities. Technological advancements further support urban transformation, with initiatives like Malaysia's Multimedia Super Corridor and Singapore's Smart Nation integrating IoT, AI, and renewable energy into urban systems. Projects in Vietnam and the Philippines, including smart traffic systems and New Clark City, showcase how technology addresses congestion and disaster resilience challenges. These efforts emphasise the need for sustainability to balance rapid urbanisation with long-term resilience.

Environmental considerations are crucial for Southeast Asia cities facing climate change and resource pressures. Kuala Lumpur and Johor Bahru invest in green infrastructure and flood mitigation, while Langkawi and Melaka focus on eco-tourism and heritage conservation. The Philippines integrates sustainable urban planning in Manila and New Clark City, emphasising green spaces and disaster resilience. Vietnam addresses diverse environmental challenges through flood mitigation in Da Nang and conservation efforts in Dong Hoi. Across the region, integrating renewable energy, green building standards, and ecological management fosters urban resilience and sustainable growth while addressing climate and resource challenges.

Governance and policies are pivotal in shaping urban development across Southeast Asia. Singapore's centralised model ensures integrated planning through initiatives like the Urban Redevelopment Authority (Henderson & Henderson, 2012), while Indonesia's decentralised system faces challenges in managing Jakarta's infrastructure. Malaysia and the Philippines emphasise regional development through initiatives like Iskandar Malaysia and New Clark City, but contend with regulatory inefficiencies and political instability. Vietnam's centralised governance under Đổi Mới fosters economic growth in cities like Ho Chi Minh City but requires inclusive strategies to address urban sprawl and housing shortages. Collaborative, adaptable governance is vital for balancing growth with equity and sustainability.

7.5 Future Directions

The challenges and headwinds of future South Asia, East Asia and Southeast Asian cities consist of several aspects, including urban developments, environmental challenges, socio-economic disparities, and political and regulatory barriers. In South Asia, countries like India, Bangladesh, and Pakistan are prioritising digital transformation, smart infrastructure, and inclusive economic growth. There is a strong push for regional connectivity through cross-border infrastructure and energy projects, alongside efforts to combat climate challenges and improve water security. Meanwhile, East Asia is focusing on advanced technologies such as AI, robotics, and quantum computing. Countries like China, Japan, and South Korea are leading in innovation, while also addressing aging populations through automation and healthcare technologies. Green energy, smart cities, and a growing role in global diplomacy are central to their future strategies. In Southeast Asia, rapid urbanisation is matched by efforts to develop smart and resilient cities, especially in Singapore, Malaysia, and Vietnam. The ASEAN block is advancing economic integration and digital trade, with sustainability high on the agenda through renewable energy and climate initiatives. The region is also investing in tourism, culture, and creative industries.

7.5.1 Tackling Urban Development

The headwinds facing urban development can be seen through the challenges to developing smart cities in Southeast Asia, including infrastructure gaps, where rapid urbanisation often outpaces the growth of necessary systems, resulting in congestion and inefficiencies. While in South Asia, cities like Amaravati, Islamabad, Dhaka, and Colombo focus on smart housing, transport, and climate-resilient planning and East Asian cities such as Songdo, Xiong'an, Tsukuba, and Taipei lead with IoT, AI, and citizen-centric innovations. Common strategies include smart mobility, green infrastructure, mixed-use zoning, and digital governance. Inequality is another significant issue, as access to innovative technologies and urban services remains uneven between urban and rural areas, exacerbating socio-economic divides. Environmental concerns, such as pollution and resource depletion, present ongoing hurdles to sustainability. Additionally, the high costs associated with implementing smart city technologies pose a significant barrier, particularly for developing economies with limited funding resources. Addressing these challenges is critical for realising the region's full potential of smart cities.

7.5.2 Tackling Environmental Challenges

Environmental challenges are among the most pressing issues affecting urban development, especially in rapidly expanding urban areas where climate change, pollution, and the need for sustainable practices present significant risks. These challenges impact infrastructure, public health, and the overall resilience of urban environments, making it crucial for planners and policymakers to address them effectively. Climate change poses significant threats to urban areas, especially those situated along coastlines and in low-lying locations. Future city programmes have been set up in Southeast Asia. The program leads 13 projects that involve nine cities and six countries. The Programme has established strategic partnerships with UN-Habitat and the UK Built Environment Advisory Group (UKBEAG), which has enabled opportunities and sustainable economic relationships between cities and international business. The countries comprising Malaysia, Indonesia, Philippines, Vietnam, Cambodia, and Thailand have taken part, and the main challenges they have responded to are environmental issues seen in multiple contexts (ESCAP, 2023).

The other example consists of rising sea levels, driven by melting polar ice and thermal expansion of the oceans, which threaten to inundate vast areas of land, particularly in cities like Ho Chi Minh City, Bangkok, and Jakarta. In addition to sea level rise, extreme weather events such as typhoons, heat waves, and heavy rainfall are becoming more frequent and severe, leading to catastrophic impacts on urban infrastructure. These events can cause widespread flooding, damage to buildings, and power outages, significantly straining the resilience of urban systems and imposing high economic costs. Urban areas must invest in resilient infrastructure, such as flood defences, green infrastructure, and improved drainage systems, to protect against these growing threats (Chen et al., 2025). In many Southeast Asian cities, inadequate waste management and a lack of strict regulatory oversight exacerbate these issues, leading to widespread waterborne diseases and environmental degradation. Addressing pollution requires a coordinated approach involving, more stringent environmental regulations, investment in waste treatment infrastructure, and public education on sustainable practices (Dao et al., 2023).

In South Asia, new urban projects like Amaravati in India and Islamabad Smart City in Pakistan emphasise renewable energy, eco-friendly buildings, and expansive green zones to reduce heat and air pollution. In Dhaka, Bangladesh, where climate risks like flooding and pollution are critical, urban planners focus on green infrastructure, improved drainage, and waste management systems. Colombo Port City in Sri Lanka includes climate-adaptive urban design and coastal protection mechanisms. Across the region, environmental policies increasingly focus on reducing emissions, conserving water, and enhancing urban biodiversity. In East Asia, cities are going further with technology integration. Songdo (South Korea) and Xiong'an (China) use IoT systems for energy efficiency, waste automation, and water conservation. Tsukuba (Japan) promotes energy-saving innovations and disaster preparedness, addressing the country's vulnerability to earthquakes and extreme weather. Taipei focuses on a circular economy, promoting low-carbon transport and citizen-driven environmental monitoring.

7.5.3 Tackling Sustainability Implementation

The struggle to implement sustainable practices amid rapid urbanisation is a pervasive challenge for many cities. The fast-paced growth of urban populations often leads to the overexploitation of natural resources, loss of green spaces, and increased carbon emissions. Urban expansion prioritises short-term economic gains over long-term sustainability, with inadequate attention given to environmental considerations. For instance, the encroachment of urban areas into wetlands and forested regions not only disrupts local ecosystems but also increases the vulnerability of cities to natural disasters such as flooding and landslides. In addressing these issues, cities are increasingly looking towards sustainable urban planning strategies that emphasise compact, transit-oriented development, green building standards, and the preservation of natural habitats. However, the transition to sustainable urban models is fraught with challenges, including financial constraints, lack of political will, and insufficient public support (Puchol-Salort et al., 2021). Moreover, successful implementation of sustainability measures requires comprehensive planning that integrates economic, social, and environmental goals and collaborative efforts among government agencies, private sector stakeholders, and communities.

Sustainability initiatives also involve promoting renewable energy, enhancing energy efficiency, and developing circular economies that minimise waste and resource consumption. Cities like Singapore and Copenhagen exemplify how integrated urban planning can successfully balance growth with environmental stewardship. Singapore's approach includes extensive green infrastructure, water-sensitive urban design, and initiatives to reduce its carbon footprint, demonstrating how cities can lead to sustainability even with limited land resources. However, replicating such models in other urban contexts, particularly in developing cities, remains challenging due to differing economic conditions, regulatory environments, and levels of public awareness. Bridging this gap will require targeted investments in green technologies, capacity-building efforts, and policies that incentivise sustainable practices.

7.5.4 Tackling Socio-economic Disparities

Socio-economic disparities are a pressing challenge in urban development, particularly in rapidly growing cities where economic growth often fails to translate into equitable benefits for all residents. Income inequality, housing crises, and the urban–rural divide are key manifestations of these disparities, impacting social cohesion and economic stability. In cities like Manila, Ho Chi Minh City, Jakarta, and Kuala Lumpur, income inequality has widened significantly due to skewed economic benefits favouring high-income groups. For instance, the wealthiest 10% in Manila controls a substantial share of the city's wealth. At the same time, low-income residents struggle to meet basic needs, a pattern echoed in other Southeast Asian cities (World Bank, 2023). Similarly, low-income workers in Ho Chi Minh City face precarious employment and limited social mobility (OECD, 2022a), while in Jakarta, the informal sector exacerbates inequalities by providing jobs without social protections. By contrast, cities like Singapore have employed comprehensive social policies, including progressive taxation and targeted subsidies, to mitigate disparities, although challenges such as cost-of-living pressures persist (Hayward, 2021).

Housing crises further deepen socio-economic disparities, with cities like Jakarta, Bangkok, and Kuala Lumpur facing soaring property prices and inadequate affordable housing. In Jakarta, a 70% surge in housing prices over the past decade has pushed many into informal settlements lacking basic amenities (ESCAP, 2023). Similar issues plague Bangkok, where average households spend over 40% of their income on housing, highlighting the strain on low-income families. Addressing these challenges requires robust policy interventions, such as expanded affordable housing and urban planning initiatives. Additionally, the urban–rural divide exacerbates disparities, as rural areas often lack access to quality infrastructure, education, and employment opportunities, fueling migration to urban centres and contributing to overcrowding and informal settlements (Das & Susantono, 2023). Initiatives like Thailand's "Smart Cities" and Malaysia's Rural Transformation Programme aim to bridge this divide by improving regional connectivity and fostering balanced development (Juasrikul & Vadenberg, 2022; Mehta et al., 2021). Sustainable urban development hinges on tailored approaches that address local governance capacities, economic contexts, and the specific needs of urban populations. Enhanced governance, progressive fiscal policies, and inclusive growth strategies are essential for reducing inequalities and fostering more equitable urban environments.

South Asia and East Asia are actively addressing economic disparities by designing inclusive urban environments that offer equitable access to opportunities, services, and resources. As both regions experience rapid surbanisation and technological advancement, there's growing recognition that economic growth must benefit all citizens—especially those historically excluded from formal systems. In South Asia, projects such as Amaravati (India) and Islamabad Smart City (Pakistan) e affordable housing, employment generation, and access to quality healthcare and education. These cities are creating inclusive zones that support small businesses, startups, and vocational training to equip marginalised communities with future-ready skills. In cities like Dhaka (Bangladesh) and Colombo (Sri Lanka), there is a concerted push to improve informal settlements, expand public transportation, and invest in digital infrastructure that connects underserved populations to the broader economy. Governments are also working to close the urban–rural divide through improved mobility, connectivity, and localised economic development. In East Asia, cities like Songdo (South Korea) and Xiong'an (China) are combining smart city technologies with inclusive planning by creating jobs in tech, green energy, and healthcare. Taipei (Taiwan) and Tsukuba (Japan) focus on digital equity, making high-speed internet and digital education accessible to low-income groups. These cities promote innovation hubs that encourage participation from youth, women, and small enterprises, ensuring broader economic participation.

7.5.5 Tackling Political and Regulatory Barriers

Political and regulatory barriers present significant challenges to urban development in Southeast Asian cities, often hindering the effectiveness of infrastructure projects and urban planning. Governance complexities, corruption, and policy gaps undermine cohesive urban strategies, particularly in cities with multi-tiered governance systems. The fragmented governance involving multiple local units also creates inefficiencies, delays, and jurisdictional disputes (Cao et al., 2021). Efforts to centralise governance, such as the Greater Jakarta Transportation Authority, often face bureaucratic hurdles and competing interests. By contrast, Singapore's centralised governance model, led by the Urban Redevelopment Authority, streamlines decision-making, integrates urban strategies and fosters sustainable growth. Initiatives like the Smart Nation program highlight how clear regulatory frameworks and centralised planning can overcome governance barriers, enabling efficient and sustainable urban development.

Corruption and policy gaps further complicate urban development, particularly in cities where weak regulatory enforcement and fragmented planning undermine sustainable growth. Corruption distorts infrastructure projects, discourages private sector investment, and inflates costs, as seen in certain Southeast Asian cities, where gaps in zoning laws, land use planning, and environmental protections lead to urban sprawl and inadequate public services (World Bank, 2023). Efforts such as Vietnam's Anti-Corruption Law and Bangkok's Climate Action Plan aim to improve transparency and policy integration but face challenges in enforcement and coordination. Singapore's integrated policy framework, exemplified by the Sustainable Singapore Blueprint, demonstrates how aligning economic, social, and environmental objectives can address urban challenges effectively. Comparative analyses reveal that overcoming political and regulatory barriers requires tailored governance structures, capacity building, and integrated planning. International initiatives, such as those by the Asian Development Bank and OECD, emphasise capacity building, transparency, and digitalisation to combat corruption and improve inter-agency coordination. Addressing these barriers through comprehensive reforms is essential to foster resilient, inclusive, and sustainable urban environments in Southeast Asia (OECD, 2024).

In South Asia, countries such as India, Bangladesh, Pakistan, and Sri Lanka face fragmented governance systems where urban planning responsibilities are spread across multiple layers of government. This lack of coordination leads to inefficiencies and delays in executing large-scale infrastructure and smart city projects. Frequent political changes and instability further disrupt long-term planning, as successive administrations often alter or abandon previous development agendas. Additionally, weak enforcement of regulations, such as zoning laws, environmental protections, and building codes, allows for unregulated growth, informal settlements, and environmental degradation. Corruption and rent-seeking behaviour in land management and procurement processes also undermine trust and transparency, skewing development priorities toward political or commercial interests rather than the public good.

While in East Asia, political and regulatory challenges are different but equally impactful. In countries like China, South Korea, Japan, and Taiwan, centralisation of authority can streamline decision-making but may limit public participation and inclusivity. For example, China's top-down planning model enables rapid development but has led to criticism over forced relocations and limited community engagement. In contrast, Japan and South Korea face issues with overly rigid bureaucratic procedures and outdated regulatory frameworks that slow innovation and infrastructure deployment. Furthermore, political tensions—such as those between China and Taiwan—can affect investment flows, international cooperation, and the implementation of cross-border smart city initiatives. There is also a persistent urban–rural divide in regulatory attention and investment.

7.5.6 Future Outlooks

Advancements in technology, sustainability, smart mobility, and inclusive development will shape the future of Asia cities. Emerging technologies such as AI, IoT, blockchain, and 5G are set to revolutionise urban management by optimising resource use, enhancing public safety, and improving governance. Initiatives like Singapore's Smart Nation program demonstrate the potential of digital integration. Still, challenges such as data privacy, cybersecurity, and the digital divide highlight the need for equitable access and digital literacy to ensure inclusive technological growth. Sustainability will also be a cornerstone, with cities increasingly adopting circular economy principles and investing in renewable energy and adaptive infrastructure to address climate change. Inclusive growth remains a priority as cities confront socio-economic disparities, including income inequality, inadequate housing, and limited public services. Investments in social infrastructure and urban planning that prioritise marginalised communities are essential for fostering economic resilience and social cohesion. Governance reform is equally critical, as fragmented systems and regulatory inefficiencies hinder sustainable urban growth.

Financing urban development is another key challenge, with the Asian Development Bank estimating Southeast Asia's annual infrastructure needs at USD 210 billion until 2030 (OECD, 2022b). Innovative funding mechanisms such as green bonds, public–private partnerships, and improved fiscal management are essential for bridging this gap. Kuala Lumpur's use of green bonds to support sustainable transport highlights the potential of such strategies. Public health resilience has also become a critical focus post-pandemic, with cities like Manila and Hanoi integrating health considerations into urban planning through expanded green spaces, air quality monitoring, and robust healthcare infrastructure; by addressing these interconnected challenges through strategic interventions, Southeast Asian cities can achieve resilient, sustainable, and inclusive growth, setting a global benchmark for urban transformation.

The future of urban development in Asia cities presents both formidable challenges and significant opportunities. As cities grow rapidly, they face intricate issues, including climate change, socio-economic disparities, governance inefficiencies, and technological integration. However, these challenges also create a unique opportunity for cities to redefine their urban planning models and embrace innovative, inclusive, and sustainable solutions. Throughout this paper, the analysis has underscored the critical barriers facing Asia cities through representative cities in South, East and Southeast Asia. Climate resilience also emerged as a crucial theme throughout the analysis, making cities increasingly vulnerable to extreme weather events, rising sea levels, and environmental degradation. Nevertheless, the effectiveness of the mitigation initiatives will depend on sustained investment, cross-sectoral collaboration, and the integration of climate adaptation into all aspects of urban planning.

7.6 Conclusion

In conclusion, we admit that future cities in Asia are at the forefront of urban transformation, combining advanced technologies, sustainable practices, and innovative planning to address the challenges of rapid urbanisation. This chapter elaborates on the current situation for Southeast countries in developing their future cities. The key drivers are urbanisation, economic growth, environmental consideration, and governance policies, the main aspects that drive and force future cities to be shaped in these regions. Asian future cities focus on smart infrastructure, green planning, climate resilience, and inclusive housing. Key trends include mobility innovation, urban tech ecosystems, digital twins, and cultural preservation. Leading cities like Singapore, Seoul, and Shenzhen pioneer sustainable, tech-driven urban models balancing smodernisation with environmental and social responsibility.

The challenges and forces for the future of cities can be seen in the context of urban development (infrastructure gaps), environmental challenges, the struggle to implement sustainability, socio-economic disparities, and political and regulatory barriers. The headwinds are created by complementing each country's various Asian statuses and conditions. Addressing these key drivers and the headwinds finally led to future outlooks on how Asia can shape their future cities. Looking ahead, the challenges facing Southeast Asian cities are multifaceted and interconnected, necessitating a holistic and adaptive approach to urban development. Achieving sustainable urban futures will require coordinated efforts from governance bodies, developers, private institutions, and civil society. Embracing integrated planning, fostering cross-sectoral collaboration, and prioritising inclusive and sustainable development will be essential. Complementing the global vision of sustainable and resilient urban approaches with embedding technological adaptation are the most popular choices among the countries. We believed that mitigating all the headwinds and standing for one nation in the region would help the countries achieve their mission of future cities.

Highlights

- Smart cities are the future focus of Asia urban development.
- The climate risks, social inequalities, strong governance and sustainability are the main keydriver and headwinds in Asia cities Focus on innovative resource management solutions to address rapid urbanisation pressures in Asia.
- Strong governance, inclusivity, and sustainability will be key to the future of Asian cities.

References

Ahsan R. & Park S. (2024). Songdo international city at the crossroads: Shifting from high-tech utopia towards reinvention of new urbanism smart city evolution: Innovation or urbanism reinvented? In *2024 International Conference of Asian-Pacific Planning Societies.* Seoul, Korea. Retrieved January 7, 2025, from https://www.researchgate.net/publication/383670196_Songdo_International_City_at_the_Crossroads_Shifting_from_High-Tech_Utopia_towards_Reinvention_of_New_Urbanism_Smart_City_Evolution_Innovation_or_Urbanism_Reinvented

Baig, Z. A., Szewczyk, P., Valli, C., Rabadia, P., Hannay, P., Chernyshev, M., Johnstone, M., Kerai, P., Ibrahim, A., Sansurooah, K., Syed, N., & Peacock, M. (2017). Future challenges for smart cities: Cyber-security and digital forensics. In *Digital Investigation* (Vol. 22, pp. 3–13). Elsevier Ltd. https://doi.org/10.1016/j.diin.2017.06.015

Bartmański, D., Kim, S., Löw, M., Pape, T., & Stollmann, J. (2023). Fabrication of space: The design of everyday life in South Korean Songdo. *Urban Studies, 60*(4), 673–695.

Bonafoni, S., Baldinelli, G., & Verducci, P. (2017). Sustainable strategies for smart cities: Analysis of the town development effect on surface urban heat island through remote sensing methodologies. *Sustainable Cities and Society, 29*, 211–218. https://doi.org/10.1016/j.scs.2016.11.005

Cao, A., Esteban, M., Valenzuela, V. P. B., Onuki, M., Takagi, H., Thao, N. D., & Tsuchiya, N. (2021). Future of Asian Deltaic Megacities under sea level rise and land subsidence: Current adaptation pathways for Tokyo, Jakarta, Manila, and Ho Chi Minh City. *Current Opinion in Environmental Sustainability, 50*, 87–97. https://doi.org/10.1016/j.cosust.2021.02.010

Chen, J., Li, H., Luo, S., Su, D., Zang, T., Kinoshita, T., & Yang, L. (2025). How do economic levels, urbanisation, and infrastructure investments influence inequality in urban green space exposure? Insights from Japanese municipalities. *Urban Forestry and Urban Greening, 104*, Article 128649. https://doi.org/10.1016/j.ufug.2024.128649

Chisty, H. U., & Aktar, M. (2022). Detailed area plan (DAP) 2022–2035 for Dhaka: The quest for identity. *Journal of Bangladesh Institute of Planners, 15*(1), 63–69. https://doi.org/10.3329/jbip.v15i1.77052

Das, A., & Susantono, B. (2023). Informal services in Asian cities: Lessons for urban planning and management from the Covid-19 pandemic. *Journal of Asian Economic Integration, 5*(2), 167–169. https://doi.org/10.1177/26316846231178601(Originalworkpublished2023)

Dao, M. S., Dang, T. H., Nguyen-Tai, T. L., Nguyen, T. B., & Dang-Nguyen, D. T. (2023). Overview of mediaeval 2022 urban air: Urban life and air pollution. *CEUR Workshop Proceedings, 3583.*

Do Livramento Gonçalves, G., Leal Filho, W., da Silva Neiva, S., Borchardt Deggau, A., de Oliveira Veras, M., Ceci, F., Andrade de Lima, M., & Salgueirinho Osório de Andrade Guerra, J. B. (2022). The impacts of the fourth industrial revolution on smart and sustainable cities. *Sustainability, 13.* Article 7165.

EC (European Commission) (2022). *Urban data platform plus; the future of cities.* Retrieved September 9, 2022, from https://urban.jrc.ec.europa.eu/thefutureofcities/urbanisation#sections

ESCAP (2023). Crisis resilient urban futures, the future of Asian and Pacific cities 2023. *United Nation 2023.* ISBN 9789210029100.

Hasan, L. (2020). *The Islamabad master plan. Policy viewpoint 16.* Pakistan Institute of Development Economics. https://doi.org/10.13140/RG.2.2.33975.55207

Hayward, D. (2021). Singapore—Context and land governance | land portal. *Landportal, March 2020*, 1–8. https://landportal.org/book/narratives/2021/singapore

Henderson, J. C., & Henderson, J. C. (2012). Planning for success: Singapore, the model city-state ? *Journal of International Affairs, 65*(2), 69–83.

Höffken, J. I., & Limmer, A. (2019). Smart and eco-cities in India and China. *Local Environment, 24*(7), 646–661. https://doi.org/10.1080/13549839.2019.1628730

Japan Counseling (2024). *Tsukuba science city analysis report.* Retrieved May 20, 2025, from https://www.counseling-japan.com/tsukuba-science-city-analysis-report/

Juasrikul, S., & Vadenberg, P. (2022). *Thailand's evolving ecosystem support for technology startups* (Issue September). https://www.adb.org/sites/default/files/publication/817496/thailand-ecosystem-support-technology-startups.pdf

Kahachi, H. A. H., Abreu, M., & Ehsan, M. (2024). Future cities' theories for sustainable future: A systematic literature review. *Futures, 164*, Article 103494. https://doi.org/10.1016/j.futures.2024.103494

Kiss, V. M. (2014). Modelling the energy system of Pécs—The first step towards a sustainable city. *Energy, 80*,. https://doi.org/10.1016/j.energy.2014.11.079

Mehta, A., Constantin, D., Crowley, S., Tirumala, R. D., Dhilion, S., Lyer, K., Andrich, M. L., & Nguyen, H. (2021). *Accelerating sustainable development after COVID-19: The role of SDG bonds* (Issue July). https://www.adb.org/publications/sustainable-development-after-covid-19-sdg-bonds

Nambu, R. & Ishibashi, Y. (2002). Effective approaches for rural development. In H. Kato (Ed.), *Approaches for systematic planning of development projects* (pp. 173–206). Tokyo: Institute for International Cooperation, Japan International Cooperation Agency.

OECD. (2022a). Development co-operation systems in southeast Asia: Indonesia, Malaysia, The Philippines, Singapore, Thailand and Viet Nam. *OECD Development Perspectives, No. 24, OECD Publishing, Paris.* https://doi.org/10.1787/50b93151-en

OECD. (2022b). G20-OECD policy toolkit to mobilise funding and financing for inclusive and quality infrastructure investment in regions and cities. *OECD Publishing, Paris*, https://doi.org/10.1787/99169ac9-en

OECD. (2024). Towards greener and more inclusive societies in southeast Asia, development centre studies. *OECD Publishing, Paris.* https://doi.org/10.1787/294ce081-en

Perez-Montenegro, C., Scanavino, M., Bloise, N., Capello, E., Guglieri, G., & Rizzo, A. (2018). A mission coordinator approach for a fleet of uavs in urban scenarios. *Transportation Research Procedia,* 35, 110–119.

Peyrard, S. &Gelézeau, V. (2020). Smart city Songdo? A digital turn on urban fabric. *Seoul Journal of Korean Studies, 33*(2), 493–518.

Potera N. (2024). The importance of surbanisation in the process of shaping the ubiquitous city on the example of Songdo International Business District, South Korea. *Studia z Polityki Publicznej, 10*, 9–24. https://doi.org/10.33119/KSzPP/2023.3.1

Puchol-Salort, P., O'Keeffe, J., van Reeuwijk, M., & Mijic, A. (2021). An urban planning sustainability framework: Systems approach to blue, green urban design. *Sustainable Cities and Society, 66*, Article 102677. https://doi.org/10.1016/j.scs.2020.102677

Rebecca G. (2022). How is Taipei becoming a Smart City? *Radio Taiwan INTL.* Retrieved May 20, 2025, from https://en.rti.org.tw/news/view/id/2008192

Schrenk, M., Popovich, V. V, Zeile, P., Elisei, P., Beyer, C., Ryser, J., Reicher, C., Carabias-Hütter, V., Musiolik, J., Lobsiger-Kägi, E., Vögeli, P., Kohler, A., & Yildirim, O. (2020). Supporting the transformation process to smart sustainable cities in Switzerland: Implementation guidelines and promising practices. In *Shaping Urban Change–Livable City Regions for the 21st Century. Proceedings of Real Corp 2020, 25th International Conference on Urban Development, Regional Planning and Information Society* (pp. 1187–1189). CORP–Competence Center of Urban and Regional Planning.

Tsoutsa, P., Panagiotakopoulos, T., Damasiotis, V., & Fitsilis, P. (2024). Empowering urban futures: Key competencies for smart city resilience officers. *Urban Governance, 4*(3), 180–192. https://doi.org/10.1016/j.ugj.2024.07.001

Tsukuba Smart City Consortium. (2024). *Articles of incorporation of the Tsukuba smart city consortium.* Retrieved May 20, 2025, from https://tsukubasmartcity.jp/en/#news

Worldatlas. (2024). *Asia map/map of Asia—Maps, facts and geography of Asia—Worldatlas.com.* World Atlas. https://www.worldatlas.com/webimage/countrys/as.htm

World Bank. (2023). *World development report 2023: Migrants, refugees, and societies.* World Bank. https://www.worldbank.org/en/publication/wdr2023

Norzailawati Mohd Noor is a professor and Deputy Dean at the Kulliyah of Architecture and Environmental Design, International Islamic University Malaysia, whose expertise lies in urban planning technologies, particularly Geographic Information Systems (GIS) and remote sensing. With a Ph.D. and M.Sc. from UTM, her research focuses on integrating advanced geospatial tools into urban planning to address challenges like urban sprawl, land use change, and disaster resilience. As Vice President of IGRSM and a registered Town Planner, she promotes the use of spatial technologies in policy-making and planning practices. Her work bridges academic research, technological innovation, and practical implementation in sustainable urban development.

Rosilawati Zainol is a professor in urban analytics at the Department of Urban and Regional Planning, Faculty of Built Environment, Universiti Malaya. She currently holds the position of Deputy Dean for Undergraduate Studies for the faculty. She specialises in urban analytics, spatial analysis, sustainable tourism planning, and walkability.

Her research interests focus on the use of information technology in urban and regional planning, particularly GIS and spatial statistics. She has published numerous articles, book chapters, and books on sustainable urban and tourism development. Some of her recent academic works include studies on sustainable tourism governance, brownfield sustainability, child-friendly city assessment tools, and sustainable transportation systems.

Part III
Future Direction of Environmental Sustainability: Adaptation and Resilience

Chapter 8
From Rising Tides to Resilient Cities: Climate Change Impacts and Challenges in Timbulsloko, Indonesia, and the Future of Other Asian Coastal Cities

Clusters of Future Studies: Corporate Foresight; Environmental Futures; Post-Normality and Complexity

Illyani Ibrahim, Hasti Widyasamratri, Mila Karmilah, Audhea Qonita, Norzailawati Mohd Noor, and M. Zainora Asmawi

Abstract This study examines the adaptive strategies employed by the community in Timbulsloko, Demak Regency, Central Java, Indonesia, to mitigate the impacts of sea-level rise resulting from climate change. It aims to share the measures taken by Timbulsloko (Indonesia) and other future cities in Asia, such as Shanghai and Guangzhou (China), Dhaka, Banishanta and Sutarkhali (Bangladesh), Ayutthaya (Thailand), and Ho Chi Minh City (Vietnam). Using qualitative methods and in-depth interviews, this study reveals how the community has developed collective behaviours, local knowledge, and indigenous technologies to adapt to permanent

I. Ibrahim (✉) · N. M. Noor · M. Zainora Asmawi
Department of Urban and Regional Planning, Kulliyyah of Architecture and Environmental Design, International Islamic University Malaysia, Gombak, Malaysia
e-mail: illyani_i@iium.edu.my

N. M. Noor
e-mail: norzailawati@iium.edu.my

M. Zainora Asmawi
e-mail: zainora@iium.edu.my

H. Widyasamratri · M. Karmilah · A. Qonita
Department of Urban and Regional Planning, Faculty of Engineering, Sultan Agung Islamic University, Semarang, Indonesia
e-mail: widya_samratri@unissula.ac.id

M. Karmilah
e-mail: mila.k@unissula.ac.id

R. Zainol et al. (eds.), *Future of Cities in Asia*,
https://doi.org/10.1007/978-3-032-01016-2_8

water inundation. These efforts demonstrate their ability to learn from past experiences and enhance their resilience in the face of environmental challenges. The findings provide valuable insights for future cities in Asia, particularly those in coastal regions that are vulnerable to the impacts of climate change. The local community in Demak regency has adopted a high level of collective behaviour to combat sea level rise flooding, which not only produces efforts but also generates local knowledge and technology as an advanced adaptation process for the future. The experience of the local community in Timbulsloko highlights the potential of hybrid adaptation models that merge traditional practices with innovative technologies. This study emphasises the importance of empowering local communities as central stakeholders in building resilience. Aligning these localised efforts with global climate frameworks, such as the Sustainable Development Goals, ensures scalability and long-term effectiveness.

Keywords Climate change · Climate resilience · Local community · Sea level rise · Timbulsloko

8.1 Introduction

This chapter examines the adaptive strategies implemented by the community in Timbulsloko in Demak Regency, Central Java, Indonesia, to address environmental challenges, while also exploring their relevance to the future of Asian cities, including those in China, Bangladesh, India, Pakistan, and Vietnam, where similar environmental issues, such as rising sea levels, flooding, and coastal erosion, have occurred, reflecting a broader global phenomenon affecting vulnerable coastal communities across various regions worldwide.

The world's largest megacities have faced significant challenges to their future sustainability due to the rapid growth of population and economic development, which affect human quality of life. According to Yeung (2001), coastal areas have experienced significant physical and socio-economic changes, resulting in serious issues related to resource utilisation and mismanagement in both their immediate surroundings and surrounding areas. Current coastal development and resource use practices are not environmentally sustainable, and coastal zones are under tremendous pressure. Water management crisis, environmental degradation, unemployment, and homelessness are among the most common urban problems in recent centuries (Nunn et al., 2021). Different climate change challenges are largely impacted in various cities, depending on their location, physical condition, population density, and the activities. Asian coastal communities account for nearly 75% of the world's total coastal population, with most residing within 10 km of the coastline (Reimann et al., 2023). Paiman and Asmawi (2017) found that coastal areas are dynamic and require technologies to accurately represent changes in shore geometry for effective management. Meanwhile, coastal environments are highly exposed to urbanisation, industrialisation, and natural hazards, which are expected to impact coastal communities (Vousdoukas et al., 2023). According to Adshead et al. (2024), climate hazards pose increasingly prolonged threats to development outcomes worldwide in coastal regions, impacting infrastructure in settlements and consequently affecting the quality of life for the community.

Coastal flooding can pose a significant issue at the local level, resulting in infrastructure damage, economic losses, and disruptions to the logistics chain. Coastal flooding also promptly affects the community's livelihood, which may not have the option to relocate. Long-term adaptation among coastal communities involves developing behaviours and enhancing the disaster resilience of coastal communities (Chang et al., 2020). Resilience, as defined by the United Nations Office for Disaster Risk Reduction (UNDRR, 2009), is the ability of a system, community, or society to resist, absorb, adapt to, and recover from hazards in a timely and efficient manner. This includes preserving and restoring essential structures and functions through risk management. Resilience involves resistance, recovery, and adaptability to various hazards such as natural disasters, climate change, pollution, and human activities while maintaining core functions and structures.

Community disaster resilience is defined as the ability of groups to seize opportunities to alter existing structures that minimise the negative impact and enhance long-term prosperity (Westman & Broto, 2022). In the socio-ecological system (SES) concept, resilience is a means of preserving and balancing between species and ecosystems to achieve landscape sustainability. Resiliency has five stages: mitigation, absorption, recovery, adaptation, and transformation. The primary key to community disaster resiliency is the ability to cope with environmental shock (Erris et al., 2022).

The Asian coastal region is highly impacted by hydrometeorological hazards that influence extreme weather, resulting in increased heavy rainfall and worse coastal flooding. According to the Asian Development Bank (ADB) (2022), sea levels have risen by an average of 21 cm since 1900, increasing at a rate of 1.7 mm per year. This rate has doubled over the past two decades, reaching up to 3.7 mm per year, and without climate policies and investments, projections suggest that sea levels could rise by up to 1 m by 2100 (ADB, 2022). Therefore, this will have a more significant impact on the coastal community.

Additionally, low-level topography and rapid land subsidence have a significant impact on the sinking cities, as well as soil compaction (Fuchs, 2010). These impacts also determine future socio-economic development, particularly in poor countries and societies. The capacity of a community to absorb, anticipate, accommodate, and rapidly recover from the consequences of a hazardous event is crucial in assessing and evaluating its resiliency. The resilience of the community, which is often hit by disasters, sometimes goes beyond understanding, as it is usually tied to subjective social factors, such as beliefs, perceptions, or attachment feelings towards their descendants. A comprehensive investigation of the unlimited and advanced community resiliency of socio-ecological systems could be enhanced in several ways by a conceptual study of disaster resilience.

8.2 Literature Review

8.2.1 Coastal Threats and Climate Change

Coastal cities face a range of threats and challenges, primarily related to environmental and human-induced factors, particularly those associated with climate change (Yeung, 2001). These challenges have significant impacts on the physical, socio-economic, and human aspects of these areas. In addition to conflicts arising from competing land uses, various global and local impacts pose threats, necessitating immediate action to mitigate their effects (Laubenstein et al., 2023). Sea level rise, exacerbated by increasing temperatures, is the most pressing global threat to coastal regions. Other significant threats include coastal erosion, flooding, storm surges, loss of biodiversity, and saltwater intrusion. These threats have profound implications for community well-being, resulting in issues such as displacement, changes in employment opportunities, and impacts on key industries, including fisheries, tourism, agriculture, and infrastructure. The vulnerability of populations and communities in coastal areas has been underscored by recent research by Xiang et al. (2024). Therefore, government interventions need to address the needs of vulnerable communities comprehensively, as failure to do so may require the active involvement of these communities in developing and implementing mitigation strategies (Zahari et al., 2013).

The number of people living in low-elevation coastal zones and their exposure to coastal flooding is highest in Asia, with notable concentrations in China, India, Bangladesh, Indonesia, and Vietnam (Adirake et al. 2021; Adnan et al., 2020; Artiningsih et al. 2016; Global Covenant of Mayors, Bangkok, Thailand 2023; Wang et al. 2018). This trend is likely to remain unchanged in the future (Neumann et al., 2015). The same problem can also seen in the Vietnamese Mekong Delta (VMD), Vietnam, which is also due to saline intrusion. Saltwater is pushing further inland through rivers and canals, contaminating freshwater sources and affecting rice fields and aquaculture, which has disrupted agriculture as a key economic sector in the region (Thach, 2023). During the 2015–2016 dry season, saltwater intrusion extended more than 90 km inland, resulting in severe crop damage in 11 of the 13 provinces within the VMD, which have made millions of people directly lost income from agricultural production, while two million people also faced shortages in drinking water and domestic water supplies due to drought and salinity intrusion (Thach, 2023).

8.2.2 *Climate Change Impact in the Other Coastal Cities in Asia*

8.2.2.1 Shanghai, China

Economic hub, Shanghai now face regular inundation threats. Meanwhile, saltwater intrusion has degraded freshwater supplies and agricultural lands in Jiangsu and Guangdong provinces, reducing crop yields by up to 20% in affected areas (Wang et al., 2023). The Yangtze River Delta, as a low-lying coastal plain, is highly vulnerable to sea-level rise and poses significant threats to both the estuarine ecosystem and urban safety in the region (Yafeng et al., 2000). Sea-level rise will cause damage to seawalls; therefore, the design standard for constructing seawalls should adopt a higher level to minimise associated risks in the Yangtze River Delta and its southern wing (Yafeng et al., 2000).

Shanghai, located in the Yangtze River Delta, faces the optimal estimated values of global sea level rise are 18 to 21 cm by 2050 (an average rate of 3.0 to 3.5 mm/a) (Yafeng et al., 2000). In China, coastal areas accommodate more than 25% of the country's population and contribute over 55% of the gross domestic product (GDP) (Wang et al., 2023). This threatens Shanghai's Pudong district, a critical economic zone, where flooding could disrupt ports, skyscrapers, and transportation networks. Economic losses are projected to reach 0.8–1.2% of Shanghai's GDP by 2100, with potential damages to infrastructure costing billions annually (Hallegatte et al., 2013).

Saltwater intrusion into the Yangtze estuary has degraded freshwater supplies, impacting drinking water and agriculture in Jiangsu province. Crop yields, particularly rice and wheat, have declined by 15–20% in affected areas due to soil salinization (Wang et al., 2023). Urban expansion has exacerbated flood risks, with impervious surfaces increasing runoff and reducing natural drainage. Since 1990, coastal flooding frequency has risen by 25%, driven by SLR, subsidence (0.5–1 cm/year), and intensified typhoons (Chen et al., 2020). Shanghai's seawalls, designed for 100-year flood events, are increasingly inadequate, necessitating upgrades to withstand 200-year events (Yafeng et al., 2000).

Adaptation measures include elevating infrastructure, expanding green spaces, and reinforcing coastal defenses. The city has invested in sponge city initiatives, such as permeable pavements and wetlands, to manage stormwater (Li, 2022). However, challenges remain, including high costs, land scarcity, and coordination between municipal and provincial governments. Integrating climate resilience into urban planning is crucial for safeguarding Shanghai's economy and its population of over 24 million (Li, 2022).

8.2.2.2 Guangzhou, China

Guangzhou is situated in the Pearl River Delta on the southeastern coast of China, characterised by complex topography that decreases from northeast to southwest

(Chen et al., 2024). According to the assessment by Li et al. (2023), approximately 36.7% of Guangdong's coastline is exceptionally vulnerable to sea level rise, rendering Guangzhou's coastal areas particularly classified as highly vulnerable to SLR, with projections of 0.3–0.6 m by 2100 under moderate scenarios (Li et al., 2023). This poses a threat to critical infrastructure, including ports and factories, which significantly contribute to China's GDP. To effectively manage these risks, immediate planning and protective measures are necessary. The necessity of strategic planning to adapt to sea level rise in coastal regions, such as Guangzhou, is emphasised through research, technical standards, policy reform, and capacity building, ensuring both environmental preservation and sustainable socioeconomic growth (He, 2012a, b).

Saltwater intrusion has reduced freshwater availability, impacting agriculture and aquaculture in the Pearl River Delta. Rice yields have dropped by 10–15% in saline-affected areas, while shrimp farming faces risks from changing water salinity (Zhang et al., 2021). Guangzhou's rapid urbanisation, with a population exceeding 18 million, has increased flood exposure by replacing natural wetlands with concrete surfaces. Storm surges, intensified by typhoons, have caused economic losses averaging $500 million annually since 2000 (He, 2012a, b).

Adaptation strategies include constructing higher seawalls, restoring mangroves, and developing early warning systems. Guangzhou has implemented the "Delta Plan," which emphasises integrated water management and green infrastructure (Wang et al., 2023). Policy reforms aim to strike a balance between economic growth and environmental protection, but challenges include funding shortages and conflicts between development and conservation objectives (He, 2012a, b). It is important to develop a comprehensive strategy for climate change, ecological civilization construction, and disaster prevention, enhance public awareness, promote data sharing, and strengthen ecological protection capabilities to enhance marine and coastal ecosystem resilience (Cai et al., 2021).

8.2.2.3 Dhaka, Bangladesh

Dhaka, Bangladesh's capital, is one of the world's most climate-vulnerable cities due to its low elevation (2–13 m above sea level), dense population (over 21 million), and socioeconomic challenges. Sea level rise (SLR) are projected at 0.3–0.7 m by 2100, combined with intensified cyclones and riverine flooding, threatens 70% of the city's land area (Hossain et al., 2022). Salinity intrusion has degraded 20% of agricultural land in coastal zones near Dhaka, reducing rice production by 15–25% and forcing farmers to adopt saline-resistant crops (Chowdhury et al., 2022). These include limited funding, unequal access to resources, institutional gaps, and the compounding effects of rapid urbanisation and poverty. Strengthening adaptive governance, equitable resource distribution, and integrating local knowledge with scientific solutions are essential to enhancing resilience and mitigating long-term climate risks (Chowdhury et al., 2022).

Rapid urbanization exacerbates flood risks, with informal settlements in low-lying areas most vulnerable. Cyclones, such as Cyclone Sidr, happened in 2007 and Cyclone Amphan in 2020, have caused billions in damages and displaced millions (Rashid et al., 2023). Adaptation measures include cyclone shelters (over 4,000 nationwide), early warning systems, and mangrove restoration along the Sundarbans, which buffers storm surges (Islam et al., 2021). However, barriers to effective adaptation include limited funding (with adaptation costs estimated at $5 billion annually), unequal access to resources, and governance gaps. Indigenous knowledge, such as floating gardens and raised homesteads, complements scientific solutions but is underutilised (Islam et al., 2021). To develop effective policies and adaptation measures against SLR impacts, it's crucial to consider community perception and evaluate different measures under different evaluation criteria, which can helps to refine adaptation measures for specific sites, disentangle SLR-induced challenges, and better plan responses to address climate change and SLR impacts (Roy et al., 2022).

8.2.2.4 Banishanta and Sutarkhali, Bangladesh

Banishanta and Sutarkhali in Bangladesh have faced land erosion and rising sea levels in the rivers and sea (Roy et al., 2022). In Bangladesh, approximately 46% of the total population lives within 10 m above sea level, and around 50 million people live below 5 m above the mean sea level (Luetz, 2018). Crop farming was the main economic activity until approximately 10–15 years ago, when the deliberate flooding of crop fields with saline water for shrimp farming, combined with erratic monsoon rainfalls and the more frequent landfall of cyclones, gradually altered the types of livelihood activities in this area (Hoque et al., 2018a, b). SLR has eroded 10–15% of arable land in these villages over the past two decades, displacing thousands and disrupting livelihoods (Luetz, 2018).

Historically reliant on crop farming, these communities shifted to shrimp aquaculture 10–15 years ago due to deliberate saline flooding and climate-induced changes, such as erratic monsoons and frequent cyclones (Hoque et al., 2018a, b). However, shrimp farming is unsustainable due to soil degradation and market volatility, which pushes residents toward migration or alternative livelihoods, such as fishing. Cyclones, occurring 2–3 times annually, cause $100–200 million in damages per event in Khulna (Ahmed et al., 2023).

Integrating local knowledge, such as traditional water management practices, with modern engineering is crucial for sustainable adaptation. The local population has developed a unique adaptation practice called 'Community Mangrove Aqua Sulviculture (CMAS)' to cultivate various floral and faunal species of the Sundarbans mangrove ecosystem (Roy et al., 2022). An adaptation also includes changing their housing structures spontaneously using their traditional knowledge, searching for new suitable jobs, migrating, emigrating, raising alternative livestock, borrowing money from relatives and microcredit lending organisations, taking national and international relief and diversifying their livelihood (Uddin et al., 2021).

8.2.2.5 Ayutthaya, Thailand

Ayutthaya, a historic city and United Nations Educational, Scientific and Cultural Organization (UNESCO) World Heritage Site in Thailand, is particularly susceptible to riverine floods, which are primarily caused by the Chao Phraya River and its tributaries. As part of the worst flooding Thailand has experienced in decades, the island, situated in the central plain of the Chao Phraya River Basin, was severely affected by flooding in the fall of 2011 (Adnan et al., 2020). Despite being seasonal, these floods have worsened due to urbanisation, upstream dam management, and climate change. The 2011 flood is the most extensively studied case and serves as a reference point for risk management in historic urban landscapes (Haraguchi & Lal, 2015).

Climate models project a 20–30% increase in monsoon rainfall intensity by 2050, amplifying flood risks (IPCC, 2021a, b). Urban sprawl has reduced natural floodplains, while the release of water from dams during heavy rains exacerbates downstream flooding. Adaptation measures include floodwalls, improved drainage, and early warning systems. The Thai government has invested $500 million in post-2011 flood mitigation, but implementation lags due to bureaucratic inefficiencies (Phongphiphat et al., 2023). The study highlights the adaptation strategies of residents to floods, including elevated residences with upper living quarters, flat-bottom boats, and pre-flood stockings. It also suggests enhancing localised flood relief efforts to better cope with repetitive flooding (Thanvisitthpon, 2019).

8.2.2.6 Ho Chi Minh, Vietnam

Can Gio District, a low-lying coastal area located southeast of Ho Chi Minh City (HCMC), Vietnam's economic hub, faces significant challenges due to climate change and sea level rise. With an average elevation of just 1.5 m above sea level, the district has experienced increasing threats to both residential and agricultural lands in recent years (Ngo & Hoang, 2020). As one of the world's most climate-vulnerable regions, its low-lying terrain (averaging just 0.8–1.2 m above sea level) makes it particularly susceptible. SLR, projected at 0.3–0.75 m by 2100, threatens 60% of HCMC's land area, home to 9 million people (Ngo & Hoang, 2020). Ngo and Hoang (2020) suggested that proposed solutions include planning residential clusters and providing architectural design models for housing projects such as stilt houses, floating houses, and lightweight concrete houses. Salinity intrusion has reduced rice yields by 20–30%, impacting Vietnam's role as a global rice exporter (Nguyen et al., 2023).

Riverbank and coastal erosion in Can Gio have displaced approximately 5,000 households since 2010, while flooding disrupts industries, resulting in annual costs of $300 million (World Bank, 2022). Adaptation includes stilt houses, floating homes, and mangrove restoration, which reduces erosion by 30% (Ngo & Hoang, 2020). HCMC's Climate Adaptation Strategy (2020–2030) emphasises resilient infrastructure and urban planning, but funding shortages and rapid urbanisation hinder progress (ADB, 2023). Integrating traditional practices, like floating markets, with modern solutions, such as flood-resistant housing, is essential. International support, including $1 billion from the Asian Development Bank (ADB), aims to bolster HCMC's resilience (ADB, 2023).

8.2.3 Linked Socio-Ecological System in Coastal Community Resiliency

The interaction between social and environmental systems has been naturally connected, blending to make it effective to assess the resiliency between humans and nature, known as the socio-ecological system (SES). Humans contribute to change, adaptation, and resilience through SES by determining the structure, building a management system, or validating the biophysical and socio-economic processes (Ruiz-Ballesteros & Paulino, 2019).

Community resiliency is created by the degree of capacity to tolerate, absorb, and cope with social or environmental changes. According to Cinner and Barnes (2019), six social factors drive social-ecological change, which are also referred to as adaptive capacity. These factors are: firstly, assets. Individuals with greater assets and access to social capital, technology, and financial resources tend to exhibit greater resilience to social-ecological changes. For example, people living in Sayung sub-district, Demak regency, who face coastal flooding in their routines, have more opportunities to move or elevate their houses. Secondly, flexibility. Flexibility represents the capacity of an institution or individual to adapt by switching strategy, technique, or methodology. Soon, flexibility is the ability to accept the consequences of any hazardous damage. Thirdly, social organisation. The society that collaborates, promotes, cooperates, accesses, and shares knowledge outside their domain might be more resilient because they can have support from their peers. Fourthly, learning. By studying the SES, people are increasing their environmental awareness. They also get some response or feedback to build or destroy their resiliency. Then, agency. To be resilient, people require power to actualise and mobilise their assets, flexibility, social organisation, and learning to scale up their future. Lastly, the socio-cognitive construct. Socio-cognitive constructs refer to people's sensitivity in processing personal experiences in hazardous events and their ability to implement social norms to minimise the risk.

Figure 8.1 (see the next page) provides the four subsystems to explain SES frameworks (Ostrom, 2009). Firstly, the resource system describes the protected area within a specified territory, including forests, aquatic systems, and wildlife. Secondly, resource units, such as types of animals, vegetation, water flow, etc. Thirdly, governance and management systems, and fourthly, users represent the individuals who utilise the environment according to specific rules, such as for research, tourism, or any other commercial purpose. This framework helps to study and identify relevant variables for designing data collection tools, preparing fieldwork, and analysing findings related to environmental sustainability.

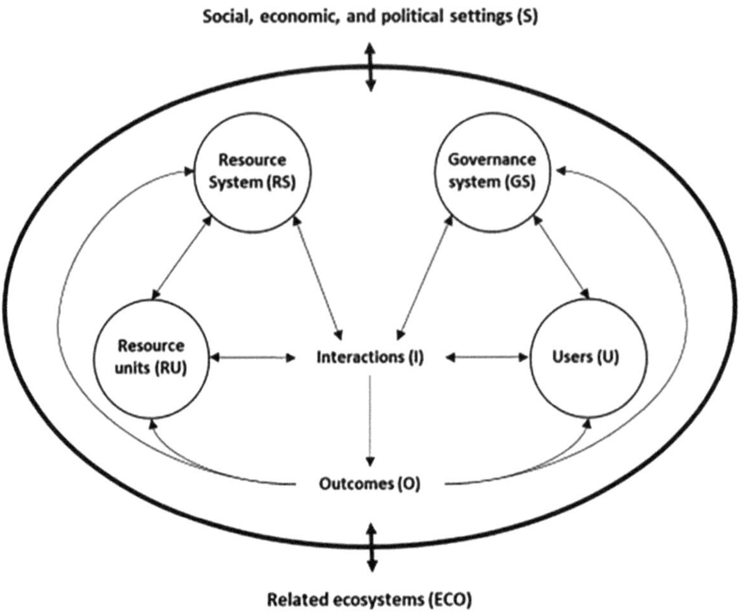

Fig. 8.1 The subsystem in the socio-ecological framework for adaptive capacity in climate change. *Source* Ostrom (2009)

8.3 Methodology

The study introduces the transformation and changes experienced by the coastal community of Timbulsloko village in Demak Regency due to climate change. The area has been affected by rising sea levels since the early 2000s, which is attributed to the development of the city of Semarang, the capital of Central Java Province. The Semarang Sea reclamation, port expansion, and Semarang-Demak Sea embankment toll road have altered the sea wave current, which in turn impacts the coastal erosion in Demak.

This study explores the adaptive strategies of the Timbulsloko community in Demak Regency, Central Java, using literature review and document analysis methods. Keywords from other Asian cities such as Shanghai and Guangzhou (China), Dhaka, Banishanta and Sutarkhali (Bangladesh), Ayutthaya (Thailand), and Ho Chi Minh City (Vietnam) and 'coastal city', 'adaptation' were used in integrated search engines and databases like Scopus and Web of Science.

By examining the transformation based on their past experiences, we can understand how the community determines its initiative and active participation to enhance its resilience capacity. At the end of the study, inferences are made that Asian coastal cities, Shanghai and Guangzhou (China), Dhaka, Banishanta and Sutarkhali (Bangladesh), Ayutthaya (Thailand), and Ho Chi Minh City (Vietnam) particularly

Table 8.1 The list of interview questions to understand the community resilience in Timbulsloko, Indonesia

Community resilience	Questions
History of the landscape changing	i. How long have you lived here? Could you share your experience related to the situation? When did the severe flooding begin? ii. Can you tell me how the water is coming in and entering your house? iii. How do you feel when water comes to your house? How much damage did you get from it?
Resiliency activity and lifestyle	i. What did you do when the water was coming? ii. When the water came, how did you go to school, the market, or any other place outside Timbulsloko? iii. How did you obtain fresh water to drink and for any other bathing activities? iv. How do you earn money for your daily livelihood? What is your occupation?
Vulnerability	i. Did you know why the water level is gradually increasing year by year and what causes this phenomenon? ii. Did you do some preparedness to anticipate this (high water inundation)? Did you monitor the weather news or read any news related to your place? iii. Did you feel stressed or frustrated living here?
Coping	i. What did you do to save your assets? ii. Are you willing to move out of the village of Timbulsloko? iii. Do you think that all of you need help or support from people or institutions other than the government? Why?
Adaptation	i. What did you do when somebody here died? How do you bury them? How is the cemetery since the Water is constantly inundating? ii. How did you fulfil your nutritional needs since the market is far away? iii. How do you solve the sanitation problem? iv. Is there a designated evacuation meeting point or public meeting area?
Short-term resilience	i. You have lived here for a long time; did you learn any lessons from this experience, particularly for the youth, to make a better life outside the village of Timbulsloko? ii. Does the village of Timbulsloko receive any attention or awareness from the government or other institutions to address or solve the problems here?

those in vulnerable regions, hold particular significance, especially those spanning extensive and densely populated coastlines.

For analysis in Timbulsloko, Indonesia, using qualitative research, we also explore how communities take their experience, knowledge, and initiative to improve their advanced resiliency. The snowball sampling technique, a non-probability method commonly used in qualitative research, was employed to conduct in-depth interviews with key individuals who have lived in the area for over 20 years, including local community leaders such as kampong leaders. To analyse the data, we categorised the data using manual coding and a triangulation data technique to check the validity.

We categorised the topics based on knowledge, experience, and initiative in order to adapt and maintain their existence as part of the community in Timbulsloko. Then, we analysed and summarised the findings of important information and innovation in their effort to enhance resiliency. Table 8.1 (see page 163) shows the list of interview questions to understand the community resilience in Timbulsloko, Indonesia.

8.4 Findings and Discussion

8.4.1 Community Social-Ecology Resilience

Located on Java Island, the most populous island in Indonesia, the northern-central Java coastal area has become a main strategic industrial hub, characterised by the magnitude of urban population activities. Meanwhile, coastal erosion and below-sea level area have forced the population to be "climate refugees", annually. According to the recorded data, fourteen sub-districts in Demak regency have been submerged (Wijaya et al., 2022), forcing the communities to relocate. One of the worst-affected sub-districts is Sayung (where Timbulsloko is one of its villages in Demak Regency, Indonesia). Sayung sub-district, one of the "climate victims," has undergone a rapid ecosystem transition since the massive urban development industry began. Over the decades, the ecosystem has undergone rapid changes. In the 1990s, the Sayung sub-district was rich in agricultural land; however, from the 2000s until the present, coastal erosion has increased due to industrial growth, driving population concentration to stay. The high-density population led to groundwater withdrawal in the coastal area, resulting in increased land subsidence rates. Figure 8.2 (see the next page) shows the timeline of changes in the coastal ecosystem of Sayung sub-district.

Figure 8.2 (see the next page) is a timeline of the coastal ecosystem transition in Sayung sub-district from the early 1990s to the present. In ancient times, the Demak regency was a coastal area located in Central Java, Indonesia, along the Muria Strait, which once separated Java from Muria Island (see Fig. 8.3 on the next page) (Bronto, 2006).The city of Demak was founded as a seaport along the Muria Strait, a trade route connecting Java and Pulau Muria, which facilitated commerce and cultural exchange (Ricklefs, 2008). This strait was active from at least the 7th to the sixteenth centuries (Reid, 1993). Meanwhile, Timbulsloko is a coastal village located in the Sayung District of Demak Regency. However, due to high sedimentation, it became land, and in modern periods, some of the locations have returned to being watery areas (Edwards et al., 1991). As a result, Timbulsloko faces significant challenges due to its coastal location.

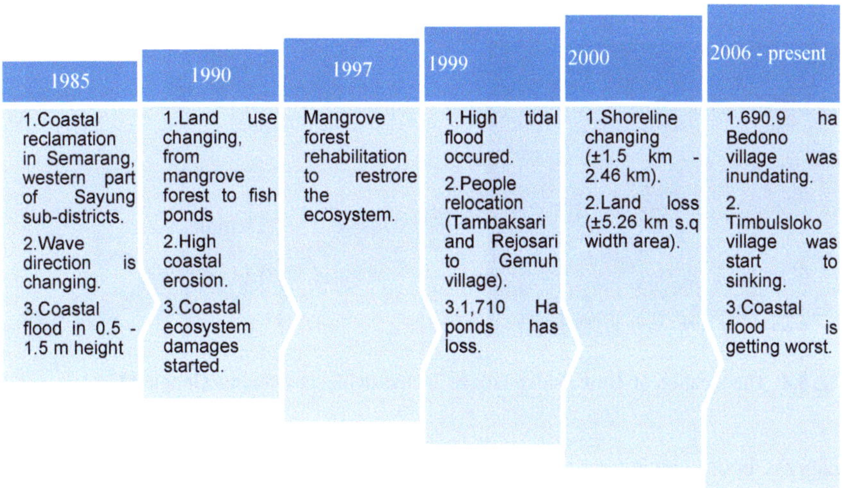

Fig. 8.2 Timeline of coastal ecosystem changes in Sayung, where Timbulsloko is one of its villages in Demak regency, Indonesia

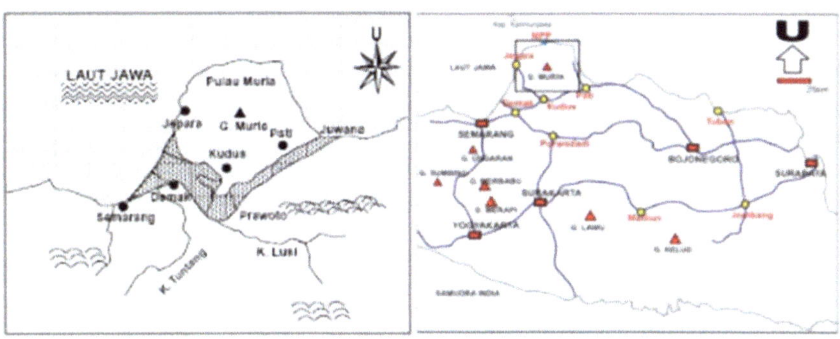

Fig. 8.3 The ancient period of the Demak regency coastal area. *Source* Bronto (2006)

By the end, scientists stated that the northern coastal region of Central Java had started to emerge. Coastal flooding is a regular occurrence and already affecting some villages in Sayung sub-districts, such as Bedono, Sriwulan, Timbulsloko, Banjarsari, Pruwosari, Gemulak, Sidogemah, Tugu, Sidorejo, and Surodadi (Asrofi et al., 2017).

In the 1990s, coastal floods occurred only once a year; recently, they have become a permanent phenomenon, entering homes through two channels: first, through slow capillaries under the floor, and second, through seawater flowing through the sewage channels. Coastal ecosystem damages impact many sectors, not only the environment, but also economic, social, and political levels, as they affect people's survival. Damage to coastal ecosystems, including mangroves, coral reefs, salt marshes, and

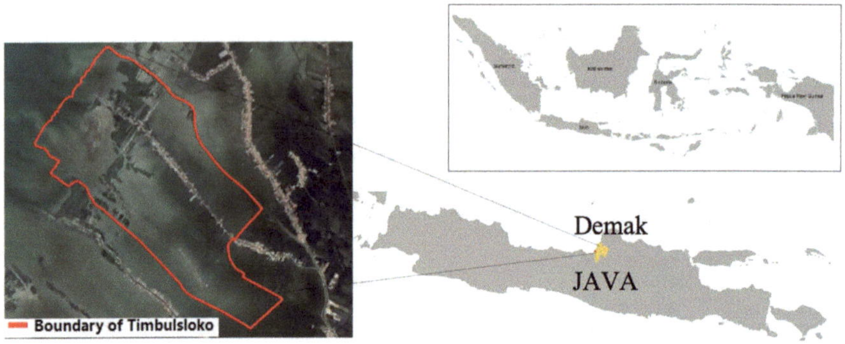

Fig. 8.4 The location of Timbulsloko village in Demak regency. *Source* Google Map

seagrass beds, can have significant and far-reaching impacts on local communities, affecting both the environment and socioeconomics. This damage increases vulnerability to natural disasters, leading to the loss of livelihoods, food and water insecurity, cultural and social disruption, and the loss of biodiversity. For residents, coastal flooding poses a serious and growing threat due to climate change. This is not a new phenomenon, as they have been facing it for years, but they have developed a high capacity for adaptation. However, assistance from the local government is crucial as it will determine the future planning and management of these sinking areas. Figure 8.4 (see above) shows the location of Timbulsloko village (indicated in red) in Demak regency (indicated in yellow).

The ecosystem change directly impacted the community resiliency pattern in the Sayung sub-district, particularly in Timbulsloko village. Social-ecological resilience is identified through people's experiences, which enable them to find new options or minimise the impact, thereby reducing risk in the future and building capacity to elevate their quality of life. Table 8.2 (see the next page) shows the social-ecology resiliency aspects for the community in Timbulsloko village, Indonesia.

Because Timbulsloko frequently experiences tidal flooding and land subsidence that damages infrastructure, including piped water systems, residents there primarily extract underground water in their homes to get freshwater due to a lack of access to a dependable public water supply. Many households are consequently forced to rely on groundwater as their primary water source because they are not connected to the local water utility. Figure 8.5 (see the next page) shows underground water extraction by the Timbulsloko's residents at their homes.

Timbulsloko community's experience of facing permanent coastal floods and their reluctance to move are creating the social-ecological 'life creativity' known as transformation, which means that the community not only needs the ability to survive but also creates new methods to be more resilient, as they would never move to another place. There are several reasons why they still live there. Firstly, economic reason. Relocating to another place incurs a high cost, in addition to the government's inability to provide sufficient compensation to these communities. Secondly, they feel an attachment to their hometown and ancestral roots, which is evident in their

Table 8.2 Social–ecology resiliency aspects for the community in Timbulsloko village, Indonesia

Socio-ecology resiliency aspects	Tools to adapt	Resiliency activity
Freshwater	Artesian well Water gallon	Sanitation Drinking water
Local transport	Boats Motorcycles	During high tides and floods Normal days
Worship	Mosque	Worship and meeting point
Education	Playgroup school	School and children Al-Qur'an recite school
Disaster mitigation	Housing	Floor lifting Wooden bridge
Local economic	Individual business	Food stall, small shop, fisherman
Social activity	Routines social activity	Disaster drill Community cooperation to maintain the Kampong wooden bridge Engaging with local traditions to stay up-to-date with the latest information

Fig. 8.5 Underground water extraction by Timbulsloko residents at their homes to get freshwater. *Source* The photograph was taken by the author, Hasti Widyasamratri, on October 9, 2022, at Timbulsloko during a site visit

Fig. 8.6 Traditional fishing and local boats as transportation in Timbulsloko, Indonesia. *Source* The photographs were taken by the author, Hasti Widyasamratri, on October 9, 2022, at Timbulsloko during a site visit

continued reliance on traditional fishing methods and local boats for daily transport. Figure 8.6 shows traditional fishing and the use of local boats for transportation.

8.4.2 Community Transformation Pathway in Timbulsloko, Indonesia

The long-term coastal floods in Timbulsloko are surely devastating the communities, as they have limited access to resources for recovery and are being forced to reach a critical point of resiliency, moving forward to transformation. The community transformation pathway was represented in four stages: (1) vulnerability, (2) coping, (3) adaptation, (4) short-term resilience. Vulnerability is closely related to the community's experience of disaster stress, physical, and economic loss. A coping strategy is a community effort aimed at coping with and surviving. Adaptation is the action taken by a community and its stakeholders to minimise risk, and short-term resilience indicates the community's ability to increase its capacity-building efforts to achieve a new direction for its future existence (Artiningsih et al., 2016; Keating et al., 2022).

Tables 8.3, 8.4, 8.5 and 8.6 explain the transformation pathways among communities in Timbulsloko village. Table 8.3 (see the next page) represents their psychological response to the hazards they face regularly. Some of them, particularly newly married women who are expected to be their husbands, experienced stress and hopelessness as they are compelled to accept this circumstance as given. However, past experiences have taught them valuable lessons. They should be prepared for various environmental challenges, many of which they had never previously encountered.

Overall, the community in Timbulsloko is gradually becoming more prepared to take the necessary steps towards adaptation and resilience.

Table 8.4 illustrates the community transformation through coping strategies, examining housing modifications, occupation changes, and social life. They built wooden bridges in front of the house for easy access. They have also learned skills to survive. The community in Timbulsloko needs to decide to protect their lives. They were divided into two groups for decision-making: firstly, the group of communities that do not have enough financial support to move or relocate to another place since they cannot sell their inundated houses. Secondly, the group of communities with better economic conditions tend to relocate to safer areas. The first group had to adapt to nature, such as making housing modifications and changing occupations, and they realised that they needed more support to find better livelihood opportunities. Thus,

Table 8.3 Stage 1 of the community transformation pathway—vulnerability: aspects of vulnerability, transformative actions and past experiences

Aspects of vulnerability	Transformative actions	Past experiences
Source of hazard	How to predict the uncertainty of coastal floods based on a particular time	Upcoming preparedness for coastal floods by monitoring the information from local government and news
Psychological	Acceptance and new beliefs to anticipate the hazard	Strengthen social unity to motivate broader community participation

Table 8.4 Stage 2 of the community transformation pathway—coping: coping strategies, transformative actions and past experiences

Coping strategies	Transformative actions	Past experiences
Housing modification	Built a wooden bridge in front of the house as access support and avoid the high tide Boats for daily activities when coastal floods are coming	The community prefers to remain in its current location rather than relocate. They choose to maintain their homes, even though the conditions are no longer suitable for living Previously, motorcycles were the primary mode of transportation when the village was still accessible by foot. However, in the current situation, residents need to move by boat
Changing occupations	People learn new skills to support their household. Previously, they were farmers; now, they were forced to be fishermen, and they work in the informal sector	The community have to get an alternative type of work
Changing social life	High dependency on support, from the neighbourhood or the government	The community needs to survive independently

Table 8.5 Stage 3 of the community transformation pathway—adaptation: adaptation strategies, transformative actions and past experiences

Adaptation strategies	Transformative actions	Past experiences
Knowledge transfer	The community has discovered a new technology to safeguard cemeteries, even in areas surrounded by water Tires surround the graveyard to avoid water intrusion Floating urban farming	Traditional cemetery (digging the soil) In the past, they lacked any knowledge or experience in urban farming
Networking	The local community started to build networks with people outside their village (academicians, NGOs, etc.)	Sharing knowledge of success evidence to fulfil better quality of life
Environmental adaptation	Rainwater harvesting to fulfil daily water Floating sanitation Fish trap	Artesian wells and traditional sanitation. Diversification of daily activity
Public space modification	Modify the mosque yard to serve as a meeting point	Promote communal engagement in future planning for a better village

Table 8.6 Stage 4 of the community transformation pathway—short-term resilience: components of short-term resilience, transformative actions and past experiences

Components of short-term resilience	Transformative actions	Past experiences
Capacity building	Community initiatives related to freshwater resources and clean sanitation, together with academics, NGOs, and local stakeholders Increase community togetherness to build the wooden bridge across the village and buy boats to support their access	Lack of environment, sanitation, and water resource awareness High dependency on local stakeholders
Social learning	The youth community is disseminating information about their village situation on social media, with the primary goal of broadening awareness about the social-ecological impact of coastal floods	Increasing public attention to get support and network from others
Livelihood transformation	New skills and opportunities to find another financial source as they switch their occupation from farming to fishing	Coastal flooding has been submerging their paddy fields, and now their paddy fields are being converted into fishponds. The local community must learn new skills to cultivate fish
Economic diversification	Women have a side job at home, where they open a food stall or mini shop and produce chips and snacks from seafood	Optimise local resources to add their income

the community in Timbulsloko is exploring new alternatives to address its skills and needs, seeking both financial and non-financial support to survive.

Table 8.5 (see page 170) illustrates the community transformation through various adaptation strategies by examining knowledge transfer, networking, environmental adaptation, and public space modification. The community demonstrated the ability to move beyond their established comfort zones, indicating a readiness for change and development. This analysis also shows that the community in Timbulsloko has begun to adopt a more open mindset, actively learning new skills and technologies to meet their daily needs. At this stage, they are increasingly supporting, encouraging, and empowering one another. They have come to understand that mere survival is insufficient; to remain visible and relevant, the community must enhance its knowledge and actively engage on social media to attract the attention of key stakeholders.

Table 8.6 (see page 170) illustrates the community transformation through the components of short-term resilience, encompassing capacity building, social learning, livelihood transformation, and economic diversification. The community is ready to take the lead in planning and building its future, although it faces limitations in terms of funding, time, and access to development agencies. The younger generation of Timbulsloko is more digitally engaged, utilising digital platforms to broaden awareness about their sinking village. They have begun to formalise the community's voice and actively participate in negotiations to improve their livelihood. By fostering interaction and collaboration, the youth are gradually developing leadership within the community.

8.5 Discussion of Implications

The case study of Timbulsloko village in Demak Regency, Central Java, Indonesia, highlights the profound impacts of climate change on coastal communities, particularly in the context of rising sea levels, coastal erosion, and land subsidence. These challenges mirror those faced by other Asian coastal cities discussed earlier, such as Shanghai and Guangzhou (China), Dhaka, Banishanta and Sutarkhali (Bangladesh), Ayutthaya (Thailand), and Ho Chi Minh City (Vietnam). By comparing Timbulsloko's social-ecological resilience and transformative pathways with these cities, we can draw broader implications for policy, community adaptation, and sustainable development in vulnerable coastal regions.

8.5.1 Common Vulnerabilities and Socioeconomic Constraints

Timbulsloko's experience with permanent tidal flooding, land subsidence, and ecosystem transition from agricultural to inundated landscapes is analogous to the challenges faced by other Asian coastal cities. For instance, Shanghai and Guangzhou (China) face similar threats from sea-level rise (SLR) and saltwater intrusion, which degrade agricultural lands and freshwater supplies (Wang et al., 2023; Li et al., 2023). In Timbulsloko, groundwater extraction due to limited public water access exacerbates subsidence, a problem also evident in Ho Chi Minh City's Can Gio District, where low elevation (1.5 m) and erosion displace communities (Ngo & Hoang, 2020).

Like Dhaka and Banishanta (Bangladesh), Timbulsloko's residents face socioeconomic barriers, including poverty and limited funding, which hinder relocation or adaptation (Chowdhury et al., 2022; Roy et al., 2022). The economic and emotional attachment to ancestral lands in Timbulsloko parallels the reluctance of Banishanta and Sutarkhali (Bangladesh) residents to relocate despite cyclone-induced losses (Hoque et al., 2018a, b). These shared vulnerabilities underscore the need for equitable resource distribution and affordable adaptation measures to support communities that cannot relocate.

8.5.2 Transformative Pathway and Community Resilience

Timbulsloko, Indonesia's four-stage transformative pathway—vulnerability, coping, adaptation, and short-term resilience—offers a framework applicable to other coastal cities. In Shanghai, adaptation includes sponge city initiatives and seawall upgrades, reflecting a shift from coping to resilience (Li et al., 2022). Similarly, Guangzhou, China's "Delta Plan" emphasises integrated water management, aligning with Timbulsloko, Indonesia's community-driven adaptations like wooden bridges and housing modifications (Wang et al., 2023).

Dhaka, Bangladesh's use of cyclone shelters and saline-resistant crops mirrors Timbulsloko, Indonesia's innovative "life creativity," such as occupational shifts and infrastructure adjustments (Islam et al., 2021). However, Timbulsloko, Indonesia's reliance on community-led initiatives due to limited government support contrasts with more resource-rich cities like Shanghai, highlighting the importance of local agency in resource-scarce settings. Banishanta and Sutarkhali, Bangladesh's shift to shrimp aquaculture, though unsustainable, reflects Timbulsloko, Indonesia's occupational transitions, suggesting that diversification of livelihoods must be supported by training and market access to ensure long-term resilience (Ahmed et al., 2023).

8.5.3 Role of Governance and Community Participation

The active participation of Timbulsloko, Indonesia's youth in using digital platforms to amplify their plight and negotiate with stakeholders is a model for enhancing adaptive governance, a need also emphasised in Dhaka (Chowdhury et al., 2022). In Ayutthaya, Thailand, community-based disaster preparedness, such as evacuation drills, supports resilience but is limited by bureaucratic inefficiencies, a challenge also faced in Timbulsloko, Indonesia, due to inadequate government compensation (Phongphiphat et al., 2023).

Ho Chi Minh City's Climate Adaptation Strategy, backed by international funding, contrasts with Timbulsloko's funding constraints, indicating that global partnerships could bridge gaps in vulnerable regions (ADB, 2023). Integrating indigenous knowledge, as seen in Timbulsloko's housing adaptations and Dhaka, Bangladesh's floating gardens, with scientific solutions can enhance resilience across these cities, but requires institutional support to scale effectively.

8.5.4 Implications for Policy and Future Directions

The Timbulsloko, Indonesia case highlights the need for tailored policies to address climate-induced displacement in coastal communities. Like Shanghai and Guangzhou, China, where urban expansion exacerbates flood risks, Timbulsloko, Indonesia challenges stem partly from Semarang's industrial development (Chen et al., 2020; Wijaya et al., 2022). Policymakers must address upstream impacts, such as sea reclamation and toll road projects, through integrated coastal zone management, as seen in Guangzhou's Delta Plan (Wang et al., 2023).

Financial mechanisms, such as climate funds or insurance, could alleviate economic barriers to relocation or adaptation, a need evident in Timbulsloko (Indonesia), Dhaka, and Banishanta (Bangladesh) (Chowdhury et al., 2022; Roy et al., 2022). Capacity building, including skill development and digital literacy, as demonstrated by Timbulsloko, Indonesia's youth, should be prioritised to empower communities in cities like Ho Chi Minh (Vitenam) and Ayutthaya (Thailand) (Ngo & Hoang, 2020; Phongphiphat et al., 2023). Finally, preserving cultural and emotional ties to ancestral lands, a factor in Timbulsloko (Indonesia) and Banishanta (Bangladesh), requires community-centred approaches that balance relocation with in-situ adaptation, such as elevated infrastructure or stilt houses.

8.5.5 Significance Beyond Timbulsloko: Insights for Other Asian Coastal Cities

Timbulsloko's transformation reflects the broader predicament of Asian coastal cities, where dense populations, industrial hubs, and low-lying geographies amplify climate risks. The case highlights the critical role of social-ecological resilience in navigating the transition from vulnerability to adaptation. While cities like Shanghai and Guangzhou, China have greater resources, Timbulsloko's in Indonesia community-driven resilience offers lessons for resource-constrained regions like Dhaka (Bangladesh) and Banishanta (Bangladesh).

Collaborative governance, equitable resource allocation, and integration of local and scientific knowledge are essential to ensure sustainable adaptation across these cities. As climate impacts intensify, Timbulsloko, Indonesia's experience, emphasises that resilience is not just about survival but about fostering innovation, agency, and hope in the face of existential threats.

8.6 Future Directions

The case of Timbulsloko (Indonesia) village in Demak Regency, Central Java, Indonesia, exemplifies the severe challenges coastal communities face due to climate change, including permanent flooding, land subsidence, and environmental degradation. These issues parallel those in Asian coastal cities, such as Shanghai and Guangzhou (China), Dhaka, Banishanta, and Sutarkhali (Bangladesh), Ayutthaya (Thailand), and Ho Chi Minh City (Vietnam), where rising sea levels, erosion, and socioeconomic constraints threaten the livelihoods of local populations. Timbulsloko, Indonesia's community has shown remarkable resilience, developing seven socio-ecological resilience aspects, freshwater management, local transportation, worship, education, disaster mitigation, regional economy, and social activity that have enabled adaptive transformations, such as elevated infrastructure and occupational shifts, mirroring strategies like saline-resistant crops in Dhaka, Bangladesh or sponge city initiatives in Shanghai, China.

Timbulsloko's (Indonesia) resilience is driven by community-led innovation, such as youth using digital platforms for advocacy, a model also seen in Dhaka (Bangladesh) and Ho Chi Minh City (Vietnam). However, limited government support hampers scalability, a challenge shared with Ayutthaya's (Thailand) bureaucratic inefficiencies and Banishanta's (Bangladesh) funding constraints. Comparative analysis reveals that while resource-rich cities like Shanghai (China) benefit from advanced infrastructure, Timbulsloko's (Indonesia) grassroots approaches offer lessons for resource-scarce regions. For instance, its shift to aquaculture aligns with Banishanta's practices, although both require training to ensure sustainability. Meanwhile, Guangzhou's (China) integrated water management provides a governance model for mitigating upstream impacts.

Robust government intervention is crucial to sustaining Timbulsloko's transformation and supporting similar communities across China, Bangladesh, Thailand, and Vietnam. This includes administrative reforms to integrate local initiatives into national plans, as seen in Guangzhou's Delta Plan; financial mechanisms, such as climate funds used in Ho Chi Minh City (Vietnam); and strategic planning to balance in-situ adaptation with relocation, preserving cultural ties, as in Banishanta (Bangladesh). Timbulsloko's experience underscores that resilience is not just about survival, but also about fostering innovation and hope, requiring collaborative governance and equitable resources to ensure sustainable adaptation across Asian coastal cities facing escalating climate risks.

8.7 Conclusion

The case study of Timbulsloko village in Demak Regency, Central Java, Indonesia, alongside comparative insights from Asian coastal cities such as Shanghai and Guangzhou (China), Dhaka, Banishanta, and Sutarkhali (Bangladesh), Ayutthaya (Thailand), and Ho Chi Minh City (Vietnam), illuminates the profound challenges and adaptive potential of communities confronting climate change. Timbulsloko's seven pillars of socio-ecological resilience—freshwater management, local transportation, worship, education, disaster mitigation, local economy, and social activity demonstrate a transformative pathway from vulnerability to resilience, paralleling strategies like Shanghai's sponge city initiatives, Dhaka's cyclone shelters, and Ho Chi Minh City's climate adaptation frameworks. These shared experiences underscore the critical role of community agency, as exemplified in Timbulsloko's youth-led digital advocacy, and the need for scalable, context-specific solutions to address common threats such as sea-level rise, coastal erosion, and socioeconomic constraints.

However, the sustainability of Timbulsloko's transformation and the resilience of comparable coastal cities hinge on robust governmental support. Administrative reforms, as exemplified by Guangzhou's integrated coastal management, financial mechanisms such as Ho Chi Minh City's international climate funds, and strategic planning that balances in-situ adaptation with managed relocation, as needed in Banishanta, are essential to overcoming resource limitations and bureaucratic inefficiencies. Timbulsloko's reliance on grassroots innovation, mirrored in resource-scarce regions like Sutarkhali and Ayutthaya, underscores the necessity of integrating indigenous knowledge with scientific approaches to foster equitable and sustainable adaptation.

Ultimately, Timbulsloko's journey offers a powerful lesson for Asian coastal communities: resilience is not merely about enduring climate impacts but about cultivating innovation, agency, and hope in the face of existential threats. By fostering collaborative governance, equitable resource allocation, and cross-city learning, policymakers can empower vulnerable communities to navigate the escalating challenges of climate change. The collective experiences of Timbulsloko and its Asian counterparts call for urgent, coordinated action to ensure a resilient, sustainable future for coastal populations across China, Bangladesh, Thailand, Vietnam, and beyond.

Highlights

- Past experiences in Timbulsloko (Indonesia), Shanghai and Guangzhou (China), Dhaka, Banishanta and Sutarkhali (Bangladesh), Ayuttaya (Thailand), and Ho Chi Minh City (Vietnam) communities drive transformation, impacting social, environmental, and economic systems.

- The community engages four stages of transformation pathway: vulnerability, coping, adaptation, and short-term resilience, reflecting responses to the pressures.
- Efforts to increase public and governmental awareness of the challenges faced the Timbulsloko (Indonesia) and other Asian coastal cities that aim to influence policy and garner support.
- Converting awareness into advocacy targets government attention and systemic support for the transformation of Timbulsloko (Indonesia) and other Asian coastal cities.
- Enhancing knowledge-sharing platforms within and beyond Timbulsloko (Indonesia) supports transformative strategies and collaboration.

References

ADB. (2023). *Financing climate resilience in Ho Chi Minh City*. Asian Development Bank Report. https://www.adb.org/sites/default/files/publication/27505/hcmc-climate-change-summary.pdf

Adirake, M., Kittitornkool, J., Armitage, D., & Chotikarn, P. (2021). An integrative approach to planning for community-based adaptation to sea-level rise in Thailand. *Ocean & Coastal Management, 212*, Article 105846. https://doi.org/10.1016/j.ocecoaman.2021.105846

Adnan, M. S. G., Abdullah, A. Y. M., Dewan, A., & Hall, J. W. (2020). The effects of changing land use and flood hazard on poverty in coastal Bangladesh. *Land Use Policy, 99*. https://doi.org/10.1016/j.landusepol.2020.104868

Adshead, D., Paszkowski, A., Gall, S. S., Peard, A. M., Gani Adnan, M. S., Verschuur, J., & Hall, J. W. (2024). Climate threats to coastal infrastructure and sustainable development outcomes. *National Climate Change, 14*, 344–352. https://doi.org/10.1038/s41558-024-01950-2

Ahmed, N., et al. (2023). Socioeconomic impacts of cyclone-induced flooding in coastal Bangladesh. *Natural Hazards, 118*(2), 1435–1456.

Artiningsih, A., Worosuprojo, S., Rijanta, R., & Hardoyo, S. R. (2016). Enhancing social-ecological resilience in Indonesia: A case of North Pekalongan District, Central Java. *Jurnal Wilayah Dan Lingkungan, 4*(3), 187. https://doi.org/10.14710/jwl.4.3.187-198.

Asian Development Bank (ADB). (2022). Sea-level change in the Pacific Islands region: A review of evidence to inform Asian development bank guidance on selecting sea-level projections for climate risk and adaptation assessments. https://library.sprep.org/sites/default/files/2022-08/sea-level-change-pacific-islands-region.pdf

Asrofi, A., Su, R. H., & Danang, S. H. (2017). Strategi adaptasi masyarakat pesisir dalam penanganan bencana banjir rob dan implikasinya terhadap ketahanan wilayah (Studi di Desa Bedono Kecamatan Sayung Kabupaten Demak Jawa Tengah). *Jurnal Ketahanan Nasional, 23*(2), 1. https://doi.org/10.22146/jkn.26257

Bronto, S. (2006). Gunung api maar di Semenanjung Muria. *Indonesian Journal on Geoscience, 2*(1), 43–54. https://doi.org/10.17014/ijog.vol2no1.20074

Cai, W., Zhang, C., Suen, H. P., Ai, S., Bai, Y., et al. (2021). The 2020 China report of the Lancet Countdown on health and climate change. *The Lancet Public Health, 6*(1), 64–81.

Chan, F. K. S., Paszkowski, A., Wang, Z., et al. (2024). Building resilience in Asian mega-deltas. *Natural Review Earth Environment, 5*, 522–537. https://doi.org/10.1038/s43017-024-00561-x

Chen, J., et al. (2020). Assessing the impacts of sea-level rise and subsidence on Shanghai's flood risk. *Journal of Coastal Research, 36*(4), 789–799.

Chen, Y., Zhang, L., & Chen, X. (2024). A framework for using event evolutionary graphs to rapidly assess the vulnerability of urban flood cascade compound disaster event networks. *Journal of Hydrology, 642.*

Chang, S. E., Yip, J. Z. K., Conger, T., Oulahen, G., Gray, E., & Marteleira, M. (2020). Explaining communities' adaptation strategies for coastal flood risk: vulnerability and institutional factors. *Journal of Flood Risk Management, 13*(4). https://doi.org/10.1111/jfr3.12646

Chowdhury, M. A., Hasan, M. K., & Ul Islam, S. L. (2022). Climate change adaptation in Bangladesh: Current practices, challenges and the way forward. *The Journal of Climate Change and Health, 6*, 1–8.

Cinner, J. E., & Barnes, M. L. (2019). Social dimensions of resilience in social-ecological systems. *One Earth, 1*(1), 51–56. https://doi.org/10.1016/j.oneear.2019.08.003

Edwards, C., Menzies, M., & Thirlwall, M. (1991). Evidence from Muriah, Indonesia, for the interplay of supra-subduction zone and intraplate processes in the genesis of potassic alkaline magmas. *Journal of Petrology, 32*(3), 555–592. https://doi.org/10.1093/petrology/32.3.555

Erris, R. F. W., Andre, N. R., & Albert N. (2022). Resilience theory: Adaptation and transformation of the film community due to the pandemic. *Formosa Journal of Social Sciences (FJSS), 1*(4), 467–484. https://doi.org/10.55927/fjss.v1i4.2218

Fuchs, R. J. (2010). Cities at risk: Asia's coastal cities in an age of climate change. *Asia Pacific Issues, 96*, 1–12.

Global Covenant of Mayors, Bangkok, Thailand. (2023). Green infrastructure for climate resiliency. https://www.asean-mayors.eu/portfolio/bangkok-thailand-green-infrastructure-for-climate-resiliency/

Haraguchi, M., & Lall, U. (2015). Flood risks and impacts: A case study of Thailand's floods in 2011 and research questions for supply chain decision making. *International Journal of Disaster Risk Reduction, 14*(3), 256–272. https://doi.org/10.1016/j.ijdrr.2014.09.005

He, X. J. (2012). Study on the sea level rising in China coast and its adaptation strategy. *Marine Forcasts.*

He, Q. (2012b). Climate change adaptation in Chinese coastal cities: The case of Guangzhou. *Urban Studies, 49*(11), 2457–2473.

Hoque, M. Z., et al. (2018a). Livelihood transitions in coastal Bangladesh: From agriculture to aquaculture. *Journal of Rural Studies, 62*, 1–12.

Hoque, S. F., Quinn, C., & Sallu, S. (2018b). Differential livelihood adaptation to social-ecological change in coastal Bangladesh. *Regional Environmental Change, 18*, 451–463. https://doi.org/10.1007/s10113-017-1213-6

Hossain, M., et al. (2022). Climate change impacts and adaptation in Dhaka: A geospatial analysis. *Climate Risk Management, 37*, Article 100432.

Islam, M. R., et al. (2021). Mangrove restoration as a climate adaptation strategy in Bangladesh. *Ecological Restoration, 39*(3), 182–192.

IPCC. (2021a). *Sixth assessment report: Physical science basis.* Cambridge University Press.

IPCC. (2021b). *Sixth assessment report: Regional impacts. Cambridge University Press.*

Hallegatte, S., et al. (2013). Future flood losses in major coastal cities. *Nature Climate Change, 3*, 802–806. https://doi.org/10.1038/nclimate1979

Keating, A., Paschen, J. A., Brown, B., Werbeloff, L., Trevor, P., Kate, F., Inara, W., et al. (2022). Transformative actions for community-led disaster resilience. Melbourne, Australia.

Laubenstein, T., Smith, T. F., Hobday, A. J., Pecl, G. T., Evans, K., Fulton, E. A., & O'Donnelll, T. (2023). Threats to Australia's oceans and coasts: A systematic review. *Ocean and Coastal Management, 231*, Article 106331. https://doi.org/10.1016/j.ocecoaman.2022.106331

Li, W., Xiang, W., Wang, H., Dong, J., Xu, H., Zhang, J., et al. (2023). Preliminary coastal vulnerability assessment for Guangdong Province, Southeast China. *Journal of Coastal Research* 39 (3), 509–518. https://doi.org/10.2112/JCOASTRES-D-21-00050.1

Li, X., et al. (2022). Sponge city construction in China: A pathway to urban flood resilience. *Water Resources Research, 58*(3).

Luetz, J. (2018). Climate change and migration in Bangladesh: Empirically derived lessons and opportunities for policy makers and practitioners. *Climate Change Management, 2018*, 59–105. https://doi.org/10.1007/978-3-319-64599-5_5

Neumann, B., Vafeidis, A. T., Zimmermann, J., & Nicholls, R. J. (2015). Future coastal population growth and exposure to sea-level rise and coastal flooding—a global assessment. *PLoS ONE, 10*, 1–34. https://doi.org/10.1371/journal.pone.0118571

Ngo, L. M., & Hoang, H. Y. (2020). Climate change and sea level rise response solutions for Can Gio District, Ho Chi Minh City. Experiences from developed countries of the world, Springer, Singapore. Editor: In: Huong L.T.T., Pomeroy G.M. (eds) AUC 2019. Advances in 21st Century Human Settlements.

Nguyen, T., et al. (2023). Climate change impacts on agriculture in the Mekong Delta. *Agricultural Systems, 206*, Article 103612.

Nunn, P. D., Smith, T. F., & Elrick-Barr, C. (2021). Path dependency and future adaptation of coastal cities: Examples from the Asia-Pacific. *Frontiers in Environmental Science, 9*, 1–14. https://doi.org/10.3389/fenvs.2021.642385

Ostrom, E. (2009). A general framework for analysing sustainability of social-ecological system. *Science, 325*, 419–422. https://doi.org/10.1016/j.ophtha.2011.11.037.

Paiman, T. and Asmawi, M. (2017). GIS application in coastal management: The perspectives of government agencies in Selangor. *Planning Malaysia, 3*, 159–170.

Phongphiphat, A., et al. (2023). Flood risk management in Thailand: Lessons from Ayutthaya. *Water International, 48*(2), 245–260.

Rashid, M. B., Sheikh, M. R., Haque, E., Patwary, M. A. A., Siddique, M. A. B., Habib, M. A., & Sarker, M. N. (2023). Consequences of catastrophic cyclone Amphan in the human-induced coastal plain ecosystems of Bangladesh. *Case Studies in Chemical and Environmental Engineering, 8*. https://doi.org/10.1016/j.cscee.2023.100467

Reid, A. (1993). *Southeast Asia in the age of commerce, 1450–1680* (vol. 2). Yale University Press

Reimann, L., Athanasios, T. V., & Lars, E. H. (2023). Population development as a driver of coastal risk: Current trends and future pathways. *Cambridge Prisms: Coastal Futures, 1*. https://doi.org/10.1017/cft.2023.3

Ricklefs, M. C. (2008). *A history of modern Indonesia since c. 1200*. Stanford University Press.

Roy, B., Penha-Lopes, G. P., Uddin, M. S., Kabir, M. H., Lucenco, T. C., & Torrejano, A. (2022). Sea level rise induced impacts on coastal areas of Bangladesh and local-led community-based adaptation. *International Journal of Disaster Risk Reduction, 73*. https://doi.org/10.1016/j.ijdrr.2022.102905

Ruiz-Ballesteros, E., & Paulino, R. B. (2019). Social-ecological resilience as practice: A household perspective from Agua Blanca (Ecuador). *Sustainability (Switzerland), 11*(20), 1–15. https://doi.org/10.3390/su11205697

Thach, K. S. R., Lee, J. Y., Ha, M. T., Cao, M. T., Nayga, R. M., & Yang, J. E. (2023). Effect of saline intrusion on rice production in the Mekong River Delta. *Heliyon 9*(10), e20367.

Thanvisitthpon, N. (2019). Adaptative strategy to mitigate impacts of repetitive flooding of residents in Thailand's Ayutthaya province. *Journal of Water and Climate Change, 10*(4), 931–937.

United Nations Office for Disaster Risk Reduction (UNDRR). (2009). Sendai framework for disaster risk reduction 2015–2030. https://www.unisdr.org/files/43291_sendaiframeworkfordrren.pdf

Uddin, M. S., Haque, C. E., Khan, M. N., Doberstein, B., & Cox, R. S. (2021). Disasters threaten livelihoods and people cope, adapt and make transformational changes: Community resilience and livelihoods reconstruction in coastal communities of Bangladesh. *International Journal Disaster Risk Reduction, 63*, 94–111.

Vousdoukas, M. I., Athanasiou, P., Giardino, A., Mentaschi, L., Stocchino, A., Kopp, R. E., Menéndez, P., Beck, M. W., Ranasinghe, R., & Feyen, L. (2023). Small island developing states under threat by rising seas even in a 1.5 °C warming world. *Nature Sustainability, 6*, 1552–1564. https://doi.org/10.1038/s41893-023-01230-5

Wang, Q. S., Pan, C. H., & Zhang, G. Z. (2018). Impact of and adaptation strategies for sea-level rise on Yangtze River Delta. *Advances in Climate Change Research, 9*, 154–160. www.keaipublishing.com/en/journals/accr/

Wang, L., et al. (2023). Integrated water management in the Pearl River Delta: Lessons from Guangzhou. *Environmental Science & Policy, 141*, 12–22.

World Bank. (2022). *Vietnam: Climate change and urban resilience.* Washington, DC.

Westman, L., & Broto, V. (2022). Urban transformations to keep all the same: The power of Ivy discourses. *Antipode, 54*(4), 1320–1343. https://doi.org/10.1111/anti.12820

Wijaya, A., Luthfi, A., Fajar, F., & Wicaksono., H. (2022). Internalisasi nilai-nilai konservasi lingkungan melalui penuturan cerita rakyat bagi anak-anak di Desa Bedono Kecamatan Sayung Kabupaten Demak. *Jurnal Abdimas, 26*(2), 222–229. https://journal.unnes.ac.id/nju/index.php/abdimas/article/view/40015

Yafeng, S., Jiwen, Z., Zhiren, X., Zixiu, J., Zixun, J. & Guishan, Y. (2000). Prediction and prevention of the impacts of sea level rise on the Yangtze River Delta and its adjacent areas. *Science in China, 43*(4).

Yeung, Y. (2001). Coastal megacities in Asia: Transformation, sustainability and management. *Ocean and Coastal Management, 44*(5–6), 319–333. https://doi.org/10.1016/S0964-5691(01)00053-9

Xiang, Z., Li, Y., & Zhang, Z. (2024). Global conservation priorities of coastal habitats towards extreme sea level rise risks. *Journal of Cleaner Production, 473*, Article 143455. https://doi.org/10.1016/j.jclepro.2024.143455

Zahari, R. K., Ariffin, R. N. R., Asmawi, M., & Ibrahim, A. N. (2013). Impacts of the 2004 Tsunami: Preparedness of the Malaysian coastal communities. In *Asia Pacific International Conference on Environment-Behaviour Studies University of Westminster, London, UK*. https://doi.org/10.1016/j.sbspro.2013

Zhang, H., et al. (2021). Impacts of saltwater intrusion on agriculture in the Pearl River Delta. *Agricultural Water Management, 245*, Article 106623.

Illyani Ibrahim is an Associate Professor and Head of Department at the Department of Urban and Regional Planning in the Kulliyyah of Architecture and Environmental Design (KAED), International Islamic University Malaysia (IIUM). She holds a Ph.D. in Applied Climate in Remote Sensing. She is a corporate member of RISM and IGRSM, and an associate member of MIP. Dr. Illyani's research focuses on applying GIS and remote sensing technologies to sustainable urban planning, land-use modeling and disaster risk assessment. She has published extensively, presented at international conferences, and collaborates with government and NGOs to promote environmental resilience and sustainable development through innovative geospatial solutions.

Hasti Widyasamratri is a lecturer at the Department of Urban and Regional Planning, Faculty of Engineering, Sultan Agung Islamic University (UNISSULA), Semarang. Her academic background includes degrees in science and engineering, with a Ph.D. focused on geospatial technologies. Her research centers on remote sensing, GIS, urban microclimates, land use change, and disaster risk mitigation. She has published studies on air temperature estimation and urban environmental analysis using satellite data. Dr. Hasti also serves on the editorial board of Pondasi, supporting research dissemination in urban and regional planning.

Mila Karmila is an Associate Professor and Chair of the Urban and Regional Planning Program at UNISSULA, Semarang. She holds a Doctorate in Architecture and Planning from Universitas Gadjah Mada. Her research focuses on tourism planning, gender studies, environmental sustainability, and disaster risk management. Dr. Mila has contributed to studies on coastal flooding vulnerability, public space dynamics, and cultural heritage preservation in flood-prone areas. She is actively involved in community-based research, emphasising resilience in urban communities and the preservation of traditional settlements. Her work integrates urban planning, environmental sustainability, and cultural heritage.

Audhea Qonita is an undergraduate student in the Urban and Regional Planning program at UNISSULA, Semarang, Indonesia. Her research focuses on climate change adaptation, particularly examining social-ecological resilience in coastal communities, as seen in her thesis on Timbulsloko village. She has also contributed to a study on using GIS to create COVID-19 resilient areas in Mangkang Kulon Village. Audhea's academic work highlights her interest in combining geospatial technology with community resilience and environmental sustainability.

Norzailawati Mohd Noor is a professor and Deputy Dean at KAED, IIUM, whose expertise lies in urban planning technologies, particularly Geographic Information Systems (GIS) and remote sensing. With a Ph.D. and M.Sc. from UTM, her research focuses on integrating advanced geospatial tools into urban planning to address challenges like urban sprawl, land use change, and disaster resilience. As Vice President of IGRSM and a registered Town Planner, she promotes the use of spatial technologies in policy-making and planning practices. Her work bridges academic research, technological innovation, and practical implementation in sustainable urban development.

M. Zainora Asmawi is a faculty member at KAED, International Islamic University Malaysia (IIUM), since 1999. She holds a degree in Urban and Regional Planning (UTM), a Master's in Environment (UPM), and a Ph.D. in Coastal Management from the University of Portsmouth, United Kingdom. A Corporate Member of MIP and a Registered Member of LPBM, she serves as Deputy Dean for Postgraduate and Research at KAED. Zainora has held several leadership roles at the university and published five books in 2017 on urban planning. Her research, funded by MOHE and INSPEN, is complemented by her active involvement in professional consultancy and community engagement.

Chapter 9
The Future of Natural Resource Conservation in Creative Cities: Insights from the Philippines, Thailand, Singapore, and China

Clusters of Future Studies: Environmental Futures; Humanity at the Limen

Jofer B. Asilum and Dayanara Bianca S. Bermudez

Abstract Banking on the creativity of the people as a major economic driver, the concept of the creative city has emerged as a promising model for development. However, the people's creativity is heavily influenced by the local culture, whose foundations include the local natural resources. The economic orientation of the creative city model could marginalise natural resources in development policy and programmes. Such marginalisation could potentially expose both natural and cultural resources to detrimental long-term harm from unsustainable development processes. Thus, this chapter utilises Ostrom's Institutional Analysis and Development Framework to analyse the institutional arrangements of creative cities and evaluate the viability of reframing the creative city model as a tool for natural resource conservation. Using document review to compare case studies of designated UNESCO creative cities in the Philippines, namely, Baguio, Cebu, and Iloilo, and their close counterparts in selected Asian countries, namely, Chiang Mai (Thailand), Singapore (a city-state), and Macao (China), respectively, this chapter examines how their governance mechanisms navigate the tension between economic development and natural resource conservation. Ultimately, the findings of this chapter offer valuable insights into how development policies and programmes can balance the economic ambitions of creative cities with the need to conserve their natural resources.

J. B. Asilum (✉)
Dr. Emilio B. Espinosa, Sr. Memorial State College of Agriculture and Technology, Mandaon 5411, Philippines
e-mail: jbasilum@alum.up.edu.ph

J. B. Asilum · D. B. S. Bermudez
School of Urban and Regional Planning, University of the Philippines Diliman, Quezon City 1101, Philippines

R. Zainol et al. (eds.), *Future of Cities in Asia*,
https://doi.org/10.1007/978-3-032-01016-2_9

Keywords Creative city · Natural resource conservation · Cultural heritage · Institutional analysis and development framework · UNESCO creative cities network

9.1 Introduction

The 'creative city' is an approach to governing cities that highlights the people's creativity as a strategic factor for sustainable development (UCCN, 2023). It is described as a place where there is a freedom of expression of one's talents, and the people's potentials are harnessed, utilised, and promoted for the common good (Landry & Hyams, 2012). Because of this, a creative city is idealised to have a well-defined, unique, and strong identity, that allows them to become an open, connected, and dynamic society welcoming innovation, entrepreneurship, and diversity (AuthentiCity, 2008). It emerged as an idea from the late 1980s as a solution to the urban decay experienced in European cities due to the effects of deindustrialisation, as industries were offshored to other countries for cheaper labour (Segovia & Hervé, 2022). Subsequently, its elaboration in Charles Landry's (2008) 'The Creative City: A Toolkit for Urban Innovators' as a cohesive policy proposal merging cultural planning and urban governance paved the way for its rising importance in the sustainable development discussion (Segovia & Hervé, 2022). In 2004, the UNESCO Creative Cities Network (UCCN, 2023) was established precisely to promote the idea of the creative city and the cooperation with and among cities around the world that use creativity as a central factor of their sustainable development policies. It covers seven creative fields, namely, Crafts and Folk Art, Design, Film, Gastronomy, Literature, Media Arts, and Music.

However, various concerns surround the creative city model, and these may be mainly attributed to the dominance of the Euro-American perspective over its global discourse (Alacovska & Gill, 2019). In Landry's (2008) operationalisation of the creative city as a policy alone, human cleverness, desires, motivations, imagination, and creativity were branded as the most important urban resources over others, including location, natural resources, and market access. While the former is supposed to be the strategic factors for the sustainable development of a creative city, for instance, in the context of developing countries outside of Europe and North America, the latter remains to have higher importance given its role in sustaining the livelihoods and alleviating the poverty and food insecurity of the low-income but usually the majority of the population (Joos-Vandewalle et al., 2018). Such an operationalisation of the creative city therefore, outright neglects the socioeconomic and geographic disparities among the world's cities, and would only lead to exacerbating the commercialisation of urban space and marginalisation of the areas that are perceived to be uncreative, resulting in increased social division and decline in a city's collective spirit (O'Connor & Shaw, 2014). However, even though the Euro-American interpretation of the creative city model predominates, it is increasingly embraced by other cities around the world (Molho et al., 2020). Therefore, prior to implementing the creative city model in diverse urban settings, it is essential to gain a comprehensive understanding of the historical background and institutional arrangements underpinning the policies that have been perceived as successful (Pratt, 2010).

A way to see whether the application of the operationalisation of the creative city model in existing creative cities in a certain context may lead to sustainable outcomes is to begin by examining the local culture. Landry (2008) explained that the creative city is 'based on the idea that culture as values, insight, a way of life and form of creative expression, represents the soil from within which creativity emerges and grows, and therefore provides the momentum for development' (p. 177). Put simply, in developing a creative city, its cultural resources serve as the raw materials to be exploited using the people's creativity as a tool (Landry, 2008). However, while culture is considered as a public good given its unlimited carrying capacity (Marrelli & Fiorentino, 2016), a local culture is highly vulnerable to the negative impacts brought about by economic globalisation and is therefore at risk of loss (Eriksen, 2021). Among the foundations of culture that make it highly vulnerable to these negative impacts is the people's relationships with nature, as natural resources serve as a basis or platform for traditional livelihoods and practices, spiritual heritage, aesthetics, and recreation, among others (UN Environment Programme, 1999). Economic globalisation in particular, necessitates the exploitation of natural resources, which may lead to their degradation or even extinction (Eriksen, 2021). Given the economic orientation of the creative city model (O'Connor & Shaw, 2014), it only aggravates such vulnerability and risk. Thus, there is a need to recalibrate the orientation of the creative city model, considering that the sustainability of a city's cultural resources rests on the sustainability of its natural resources (Eriksen, 2021). This further implies that an analysis of the institutional arrangements surrounding a creative city's natural resources is necessary to ensure their sustainability (Ostrom, 2011), and eventually lay the foundation of the sustainability of its cultural resources.

This chapter argues that the sustainability of a creative city's natural resources is a requirement for its sustainable development. Thus, natural resources should not be placed under but alongside human creativity as a strategic factor for its sustainable development. It aims to propose a conceptual framework that integrates natural resource conservation in the creative city model. This shall be built by: 1) explaining the relationships between the creative city model and natural resource conservation and 2) by analysing the institutional arrangements allowing or hindering the UNESCO creative cities in the Philippines, namely, Baguio, Cebu, and Iloilo, and their close counterparts in selected Asian countries, namely, Chiang Mai (Thailand), Singapore (a city-state), and Macao (China), to address the tensions between the promise of economic development as creative cities and natural resource conservation.

9.2 Literature Review

9.2.1 Creativity and the City

The city maintains an important role in local and regional development despite its evolution over the centuries, specifically on its size, form, structure and composition as it now becomes a positive force in addressing sustainable economic growth, development, and prosperity (UN Habitat, 2020). As cities concentrate, accelerate, and diversify, social and economic activity also increases (Bettencourt & West, 2017), since cities tend to become more creative. Moreover, creativity is a stimulus for innovation, sustainability, and liveability in urban planning. By embracing creative approaches, cities can unlock their full potential as dynamic, inclusive, and thriving spaces for residents, businesses, and visitors.

One of the ways creativity shines in urban planning is through innovative urban design strategies. Decision makers within the city adopt creative design concepts to reimagine public spaces, streetscapes, and buildings, making them more walkable, accessible, and conducive to social interaction. Ideas like mixed-use developments, green spaces, and pedestrian-friendly infrastructure not only make cities visually appealing but also contribute to healthier and more inclusive urban environments (Gehl, 2012). Furthermore, the concept of the "creative city" emphasises the importance of honing creativity and cultural expression as factors of urban development. Creative centres, cultural institutions, and arts-based initiatives contribute to the distinct identity of cities, fostering a sense of place and belonging among residents (Landry, 2008). These creative assets not only attract tourists and visitors but also enhance the social and economic vitality of urban communities.

9.2.2 The Making of a Creative City

In transforming a city into a place that creatively tackles sustainable development challenges, Landry (2008) presents a participatory five-phase process. The first phase identifies the problems and aspirations of the city, and the needs, including the projects, stakeholders, partners, and political support, to address them in creative city terms. The second phase audits the local internal and external resources of a city. It also assesses their potential and obstacles in sustaining the cycle of urban creativity which is at the centre of addressing the city's problems and aspirations in creative city terms. The cycle involves 'enhancing ideas-generating capacity, turning ideas into practice, networking and circulating, platforms for delivery, *and* building markets and audiences' (Landry, 2008, p. 225) to generate new ideas. The third phase identifies the indicators of how far the city has gone in kick-starting its cycle of urban creativity and of the projects to address the gaps that hinder it from doing so. The fourth phase is the implementation of the said projects and the necessary organisational structure to sustain the cycle. Lastly, the fifth phase communicates and disseminates the results of the cycle, and reflects upon them for improvements.

To aid the creative city making process, Landry and Hyams (2012) identified ten key indicators of a creative place, namely, (1) political and public framework (2) distinctiveness, diversity, vitality, and expression, (3) openness, trust, accessibility, and participation, (4) entrepreneurship, exploration and innovation, (5) strategic leadership, agility, and vision, (6) talent development and the learning landscape, (7) communication, connectivity, and networking, (8) the place and place-making, (9) liveability and well-being, and (10) professionalism and effectiveness. The UCCN (2023) adopts these indicators and Landry's (2008) participatory five-phase creative city making process and elects new members based on how the applicant cities have actualised them. Furthermore, the European Union also builds on these indicators and created the Cultural and Creative Cities Monitor, which assesses the cultural and creative liveliness of European cities based on their cultural vibrancy, creative economy, and enabling environment (Alberti et al., 2023).

9.2.3 *The Conservation of Natural Resources*

Landry and Hyams' (2012) indicators, and the subsequent applications of the UCCN (2023) and the European Union (Alberti et al., 2023), reaffirm the integral role of cultural resources in creative city making, wherein a local culture influences the manner its people develop and express their creativity (Cabra & Guerrero, 2022). However, a local culture is not without its people's interactions with their natural surroundings (UN Environment Programme, 1999). Mason (1896, 1905), as cited in Freilich (1967), noted that environmental factors, including physical geography, climate, predominant flora and fauna, and mineral resources, play a significant role in shaping cultural development. This emphasises the importance of the conservation of natural resources in sustaining a city's cultural resources (Eriksen, 2021). Natural resource conservation is the management of the human uses of the natural environment in ways that enable it to produce the greatest sustainable benefit to the present and future generations, including the preservation, maintenance, sustainable use, restoration, or enhancement of natural ecosystems (IUCN, 1980).

Margules and Pressey (2000) discuss the six-stage process for planning natural resource conservation. Initially, the process entails gathering information regarding the biodiversity of a given area, especially the rare and threatened species. Subsequently, conservation objectives are identified in the second stage, encompassing both quantitative metrics such as the minimum size and connectivity of conservation areas, as well as qualitative goals aimed at mitigating human-induced disturbances. The third stage involves evaluation of existing conservation areas, assessing their efficacy in meeting targets while also identifying emerging threats. Following this assessment, the fourth stage entails the selection of new conservation areas that will contribute to attain regional conservation goals, considering existing conservation programmes, budgetary constraints, and opportunity costs for other land uses, among others. Implementation of conservation activities occurs in the fifth stage, aimed to achieve the identified targets. Lastly, the sixth stage focuses on the ongoing maintenance of the required values of conservation areas by monitoring the key indicators of conservation targets.

9.2.4 *Analysing the Creative City's Resource Conservation Policy*

Using a city's creativity in attaining its resource conservation objectives requires an understanding of the institutional arrangements that enable or hinder it (Ostrom, 2011). One tool to do so is Nobel laureate Elinor Ostrom's Institutional Analysis and Development Framework (Polski & Ostrom, 2017) (Fig. 9.1). Central to this framework is analysing the action arena, specifically the behaviour of individuals and groups (or actors) consistently involved in an action situation. Three external variables influence the actors' behaviour in an action situation, namely, the local biophysical conditions, the local community attributes or culture, and the rules-in-use (i.e., actions that are required, prohibited, or permitted). On the other hand, the actors' behaviour in an action situation results in patterns of interaction whose outcomes contribute or hinder the economic efficiency, equity through fiscal equivalence, redistributional equity, accountability, conformance to values of local actors, and sustainability of resource use.

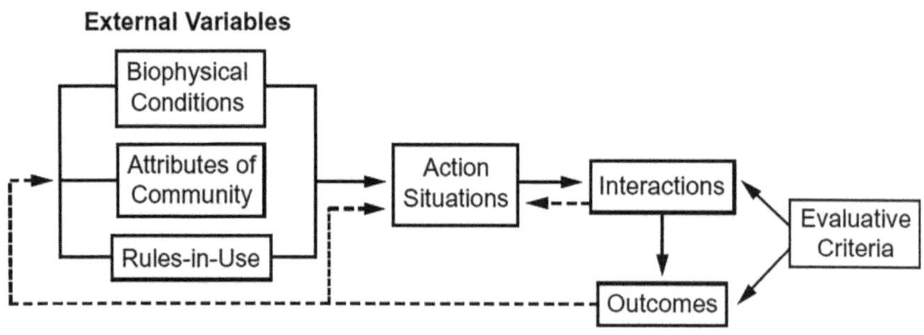

Fig. 9.1 Institutional analysis and development framework (adapted from Ostrom, 2011)

9.3 Methodology

Applying the Institutional Analysis and Development Framework, this chapter asserts that the institutional arrangements necessary for a successful resource conservation policy in a creative city mirror the same markers and procedures that define a creative locale. Drawing on the case studies of designated UNESCO creative cities in the Philippines and their close counterparts in selected Asian countries for comparison, this chapter examines how their governance mechanisms navigate the tension between economic development and natural resource conservation. The focus is on the Philippines, given its status as a developing country actively working on city development efforts eligible for UCCN recognition through the enactment of the Philippine Creative Industries Development Act (2022). However, its national policy, which primarily incorporates the operationalisation of the creative city model in the Euro-American context, overlooks the significance of natural resources in the creation and preservation of cultural resources. As one of the world's seventeen megadiverse countries possessing the majority of the world's species and high degrees of endemism (Keong, 2015), the Philippines faces vulnerability to the adverse effects of unsustainable economic practices reliant on natural resources. It has three cities elected to the UCCN, namely, Baguio in 2017 for Crafts and Folk Art, Cebu in 2019 for Design, and Iloilo in 2023 for Gastronomy.

A comprehensive review of relevant policies, news articles, and research papers[1] was conducted to identify and analyse the institutional arrangements surrounding resource conservation in the Philippines' three creative cities and their close counterparts in selected Asian countries, namely, (1) Chiang Mai in Thailand (elected to the UCCN in 2017 for Crafts and Folk Art) for Baguio; (2) Singapore (a city-state elected to the UCCN in 2015 for Design) for Cebu; and (3) Macao in China (elected to the UCCN in 2017 for Gastronomy) for Iloilo. Document review was chosen for the case study for its effectiveness in efficiently gathering and analysing diverse information concerning how these creative cities balance their economic requirements alongside natural resource conservation.

9.4 Findings

Table 9.1 (see the next page) focuses on Baguio (Philippines) and Chiang Mai (Thailand), both located in mountainous regions and highlights their artisanal heritage in their respective creative economies. On the other hand, Table 9.2 (see page 191) examines Cebu (Philippines) and Singapore (a city-state), both cosmopolitan port cities fostering design thinking and innovation to address urban challenges. Lastly, Table 9.3 (see page 193) analyses Iloilo (Philippines) and Macao (China), both showcase their culinary heritage and the influence of their colonial pasts as important trading ports.

[1] Reviewed literature for Baguio (Philippines) includes de Guzman (2020), Santos (2020), and Castillo (2023); for Cebu (Phillippines): World Bank Group (2017), Montera and Lin (2018), Chuangchai (2020), and Elizaga and Chua (2021); for Iloilo (Philippines): Oabel (2003), Iloilo City 2023–2028 Comprehensive Development Plan (2023), and Salas (2023); for Chiang Mai (Thailand): Chuangchai (2020), Pongruengkiat et al. (2023), and Rinchumphu et al. (2024); for Singapore (a city-state): Design 2025 Master Plan (2015) and Chaudhury and Lundberg (2018);

Table 9.1 Institutional arrangements of resource conservation in Baguio (Philippines) and Chiang Mai (Thailand)

Institutional analysis and development domain	Baguio (Philippines)	Chiang Mai (Thailand)
Physical world	A plateau city with a cold climate and surrounded by forested mountains	A city in the Thai highlands with fertile valleys defined by the Ping river
Community	Cordillera mountain range as the foundation of the Igorot indigenous people's culture	Mount Suthep as an important icon of the people's Buddhist traditions
	Crafts and folk art (e.g., silver crafts, basketry, weaving, wood carving, tattooing) are heavily inspired by the indigenous culture	Crafts and folk art (e.g., pottery, silverwork, silk embroidery, lacquerware) are heavily inspired by the indigenous culture
	Local city dwellers fuel the creative sectors of crafts and arts	Local civil society leads the development of crafts and folk art
Rules-in-use	Local creative city policy functions mainly as a complement to the tourism industry	Two existing creative city policy orientations: creative Chiang Mai focuses on technology-based programs, aiming to make the city more attractive for investment and claims to be volunteer-based, while Chiang Mai city of crafts and folk art leads a culturally sensitive approach to city growth and focuses on crafts development, although follows a top-down management approach
	Local cultural resources as economic resources to support the creative economy	Local cultural resources as economic resources to support the creative economy
	Land use plan covers 2024 to 2032	Comprehensive city plan crafted in 2021
Action arena:		
Actors	Local government unit, creative sector, national government agencies	Local government organisations, universities, international organisations and institutions, creative sector
Action situation	Maintenance of natural resources to preserve Igorot culture	Maintenance of natural resources to preserve Lan Na culture

(continued)

Table 9.1 (continued)

Institutional analysis and development domain	Baguio (Philippines)	Chiang Mai (Thailand)
Patterns of interaction	City governance is based on the vision to be a creative, inclusive, and liveable city	City governance is based on the vision to be country's 'most splendid city of culture'
	Cultural elements are employed to promote tourism	Cultural elements are employed to generate investment, although approached differently by actors leading the competing policy orientations
	Land use policy implementation exhibits a strong neoliberal stance and is deemed unsustainable	Ineffective implementation of land use policies, despite featuring a forward-thinking urban planning approach aiming to sustain natural and cultural resources
Outcomes:		
Economic efficiency (cost effectiveness)	Culture and tourism are highly integrated as economic engines	Craft industry remains low in economic value and on the hierarchy of policy priorities, with government programs focusing on technological development
Equity through fiscal equivalence (derive greater benefits pay more)	Local civil society is highly involved in the creative economy	Local civil society is highly involved in the creative economy
	Economic investments take precedence over sustainable tourism	Economic investments for craft industry to increase community benefits
Redistributional equity (individuals contribute based on their capacity)	Recent government recognition and support for creative workers delay the sustainability of livelihoods	Actors of urban change and culture have little cooperation amidst competing policy orientations resulting in overlapping cultural initiatives
	Diminishing sense of belonging to locals	
Accountability	Local policy institutionalises the creative Baguio City council to manage and develop the local creative community	Governor of Chiang Mai-led creative Chiang Mai's projcets are not directly funded by provincial government offices, while Chiang Mai Provincial Administration Organisation (PAO)- and Chiang Mai University (CMU) Faculty of Fine Arts-led Chiang Mai City of Crafts and Folk Art's are funded by the PAO and CMU

(continued)

Table 9.1 (continued)

Institutional analysis and development domain	Baguio (Philippines)	Chiang Mai (Thailand)
Conformance to values	Creative city status aggravated cultural appropriation contributing to cultural degradation	Craft industry developed to supply external demand and does not conform to local traditions
Sustainability of resource use	Predominantly commercial orientation of the local creative and tourism sectors poses challenges to environmental conservation efforts	Lack of coordination among actors endangers sustainability of cultural resources and risks further detachment from said resources' roots in nature

The tables illustrate the significant impact of local policies on shaping a creative city, influencing human behaviour, governance, and developmental outcomes, thereby guiding conservation efforts. Mentioned cities above derive much of their cultural heritage from the interactions with the physical world. Additionally, the rules-in-use of these cities play an important role in shaping directions towards sustainable development. For instance, Baguio (Philippines), despite efforts towards sustainable tourism, its policies towards local creative and tourism sectors are highly commercial, posing challenges to environmental conservation. Its close Asian counterpart, Chiang Mai (Thailand), amidst its competing policy orientations on creative city processes, struggles to ensure collaboration among its key actors, risking the sustainability of its cultural and natural resources.

Meanwhile, Cebu (Philippines) has fostered a vibrant community of creative entrepreneurs but struggles to align policies with best practices for environmental conservation due to its focus on being globally competitive. In the case of Singapore (a city-state), which perhaps exhibits a potential future of Cebu (Philippines), showcases how a heavily planned globalised economy results in growth, the enhancement of urban biodiversity, and a high quality of life, but may marginalise free and spontaneous creative activities, particularly happening on the ground. Both cases nevertheless highlight the negative impacts of economic globalisation on local cultures (Eriksen, 2021) particularly when not given due consideration, as evidenced by the lack of emphasis on nature-culture interactions on the creative economy.

Conversely, Iloilo (Philippines) has focused on leveraging its natural resources for urban development, resulting in a robust foundation for its identity as a creative city, balancing economic benefits with environmental sustainability. Despite its recent establishment and on-going implementation, Iloilo (Philippines) is progressing in its journey as a creative city, emphasising the need for continuous attention to maintain

and for Macao (China): Macao Special Administrative Region (SAR) Tourism Government Office (2021), Huang (2022), Wu et al. (2022), Jackson (2023), Macao SAR (2023).

Table 9.2 Institutional arrangements of resource conservation in Cebu (Philippines) and Singapore (a city-state)

Institutional analysis and development domain	Cebu (Philippines)	Singapore (a city-state)
Physical world	A coastal city with rugged and mountainous areas serving as the Philippines' second port	An island city-state serving as an important global port situated in the main shipping channel between the Indian and Pacific oceans
Community	Collective identity is strongly influenced by Spanish tradition i.e., religiosity as reflected by the celebration of the Sinulog, a festival showcasing devotion to Santo Niño, a religious image of the Child Jesus	Collective perception of city-state as a nation by design, or that is heavily planned
	Eurocentric (colonial) art is overwhelming folk practices and contemporary art forms	Design is valued beyond aesthetics and is used to co-create better living environments and a stronger national identity
	Sinulog is originally a precolonial prayer-dance ritual mimicking the river current and honouring local deities and ancestors.	
Rules-in-use	Local creative city policy is patterned after Western creative cities and is a strategy for rapid economic growth	Creative policy promotes constant reinvention of the nation to meet global demands but emulates Western creative cities
	Land use plan covers 2023 to 2032	Presence of a design master plan for an envisioned Singapore in 2025
Action arena:		
Actors	Local government units, creative sector, national government agencies	National government organisations, creative sector
Action situation	Maintenance of natural resources that serve as basis for the precolonial meaning of Sinulog	Enhancement of natural resources to ensure a high quality of life
Patterns of interaction	City governance is guided by the vision to be globally competitive and an ecologically balanced city and by its status as the current 'Queen City' of southern Philippines	Highly corporate city governance is based on the vision to make Singapore an innovation-driven economy and a lovable global city by design

(continued)

Table 9.2 (continued)

Institutional analysis and development domain	Cebu (Philippines)	Singapore (a city-state)
	Design industries are used to attract investors	Design industries function as key drivers of innovation, value creation for businesses, solutions to societal problems, and nation-building
Outcomes:		
Economic efficiency (cost effectiveness)	Culture and urban economy are disconnected	Design as an engine for innovation-led economic growth
Equity through fiscal equivalence (derive greater benefits pay more)	Creative sector usually engages specialist agencies and professionals	Design is infused into the national skillset to hone future generations
	Economic expansion due to global market engagement	Creative industries as a platform to maintain global relevance
Redistributional equity (individuals contribute based on their capacity)	Creative sector's focus on specialist entities resulted in lack of engagement of local communities	Creative city processes' focus on promoting sense of belonging and ownership, although highly regulated (e.g., Instagram-worthy graffiti recently considered acceptable wall art by authorities)
	Little to no progress in the industry as there were no alternate intervention to challenges	
Accountability	No local policy institutionalising creative city processes	DesignSingapore council established in 2003 under the then Ministry of Information and the Arts to manage creative city processes
	Creative sector organisations usually do not develop in the long-term resulting in insufficient mechanisms to ensure accountability	Series of government plans implemented transforming the city-state into a creative city (i.e., Renaissance City Plan for 2000-2015, Design 2025 Master Plan for 2015-2025)
Conformance to values	Creative activities lack understanding of social diversity	Design activities are a key component of national identity, although cultural production is generally setup to become as vibrant as Western global cities, e.g., London, New York
Sustainability of resource use	Outdated land use plan and disconnected creative city processes contribute to unregulated spatial development	Holistic approach of master plan ensures design is used to enhance biodiversity and address other environmental challenges

Table 9.3 Institutional arrangements of resource conservation in Iloilo (Philippines) and Macao (China)

Institutional analysis and development domain	Iloilo (Philippines)	Macao (China)
Physical world	Coastal city with vast plains and with the Iloilo River defining its landscape	City on reclaimed land, that is relatively flat and is at the centre of the Guangdong-Hong Kong-Macao greater bay area megalopolis, surrounded by the Lingdingyang Channel
Community	Local identity is heavily influenced by the city's past as the 'Queen City' of southern Philippines and the centre of the national sugar industry during the Spanish colonial era	Heavily influenced by its history as a former Portuguese colony, its significance as a key international trade hub, and its enduring Chinese cultural heritage
	Signature local dishes reflect the city's local ingredients and its strong affinity to Chinese and Spanish cultures due to its colonial role as a global trading port	Macao's culinary traditions reflect the city's diverse community of Chinese, Portuguese descendants, and international residents, serving as a source of local pride and playing a vital role in its vibrant cultural identity, tourism, and creative economy
Rules-in-use	Land use policy is focused on the development of the Iloilo River Esplanade Network (waterfront-oriented development), including greening of smaller waterways, and revitalisation and reuse as business centres of heritage districts	Land use policies prioritise the development of cultural districts by revitalising heritage sites and strengthening the local gastronomic identity through strategic urban planning
	Strong public-private collaboration is promoted in creative city processes	Strong public-private collaboration is promoted in creative city processes
	Land use plan covers 2021 to 2029	Urban Master Plan covers 2020 to 2040
		'Deepen Construction of Creative City of Gastronomy' is included in the Macao Tourism Industry Development Plan
Action arena:		

(continued)

Table 9.3 (continued)

Institutional analysis and development domain	Iloilo (Philippines)	Macao (China)
Actors	Local government units, creative sector, national government agencies	Local government, creative sector, national government agencies, private sector, international organisation
Action situation	Maintenance of natural resources necessary for cultural heritage preservation, especially colonial infrastructure	Maintenance of natural resources to preserve culinary traditions while modernising the gastronomic industry and balancing tourism growth with the conservation of Portuguese and Chinese heritage
Patterns of interaction	City governance and innovation are guided by the principles of natural resource conservation and cultural heritage preservation while sustaining a robust, inclusive, and resilient economy	City governance emphasises collaboration among public authorities, the private sector, and community stakeholders, particularly in the realms of tourism and gastronomy
Outcomes:		
Economic efficiency (cost effectiveness)	Culture, natural environment, and urban economy are interconnected	Design plays a role in driving economic growth, with ongoing efforts to align it closely with sustainable development goals
Equity through fiscal equivalence (derive greater benefits pay more)	Creative sector is empowered through strong government and private sector support	Creative sector is empowered through strong government and private sector support
Redistributional equity (individuals contribute based on their capacity)	Needs of the creative community are supported by public and private sectors	Creative sectors from diverse backgrounds are supported by the government
Accountability	No local policy institutionalising creative city processes yet, but the city government and the private sector collaborate in managing and developing the creative economy	No local policy institutionalising creative city processes yet, but the city government, cultural organisation, and the private sector collaborate in promoting creative economy

(continued)

Table 9.3 (continued)

Institutional analysis and development domain	Iloilo (Philippines)	Macao (China)
Conformance to values	Creative city status strengthens local cuisine and its use of indigenous ingredients	Creative city status strengthens local cuisine and its use of indigenous ingredients
Sustainability of resource use	Orientation of the local land use policy and creative city processes on heritage preservation contributes to improving the city's environmental conditions	Land use policies and creative city initiatives aim to foster sustainable resource use while prioritising the preservation of the city's natural and cultural heritage

its cultural and environmental integrity. Its fellow gastronomy city, Macao (China) has placed significant emphasis on preserving its cultural heritage by enhancing its gastronomic identity. This commitment is reflected in its development plans, which prioritise its status as a creative city through collaborations with the private sector and cultural organisations, while also ensuring environmental sustainability, similar to Iloilo (Philippines). In fact, Macao (China) is progressing as a creative city by promoting the sustainable development of the local gastronomic culture industry (Macao SAR Government Tourism Office, 2021).

The reviewed literature and the institutional analysis highlight that the concept of the creative city can be and needs to be expanded to integrate natural resource conservation. In essence, Fig. 9.2 (see the next page) shows that a city's cultural resources are based on its local natural resources, whose conservation mechanisms include preservation, maintenance, sustainable use, restoration, and enhancement. For cultural resources to become the foundation of the local people's creative solutions in addressing sustainable development challenges, Landry and Hyams' (2012) indicators of a creative place must be fulfilled to commence and sustain the people's creativity cycle. Such solutions can be utilised to enhance the city's conservation mechanisms to also enhance and sustain its natural resources. On the other hand, Fig. 9.3 (see the next page) applies the expanded conceptual framework in the creative city planning framework. In line with attempts on making the creative city model more inclusive (Alacovska & Gill, 2019), to mainstream resource conservation planning, resource conservation must be immediately identified as a challenge and aspiration of the city in the first phase of the creative city planning process. From here, each phase of the resource conservation planning process is mainstreamed in each of the rest of the phases of the creative city planning process.

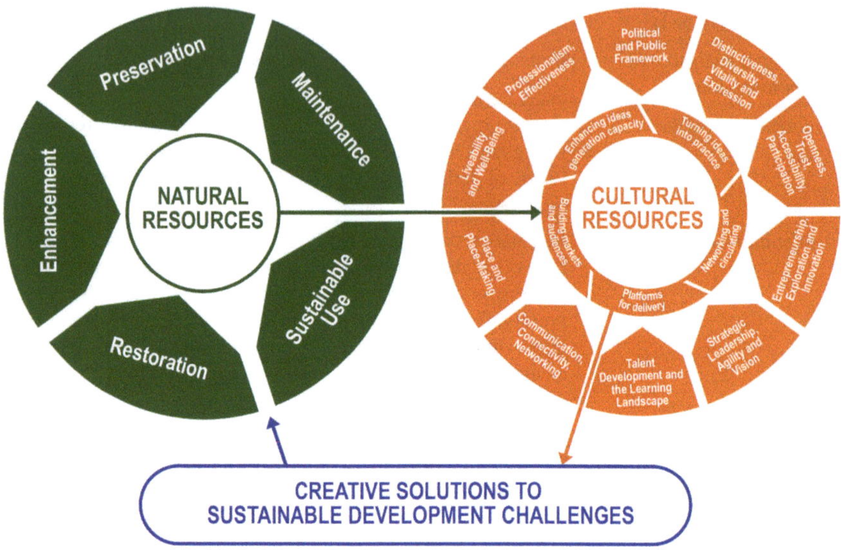

Fig. 9.2 Integration of natural resource conservation in the creative city model

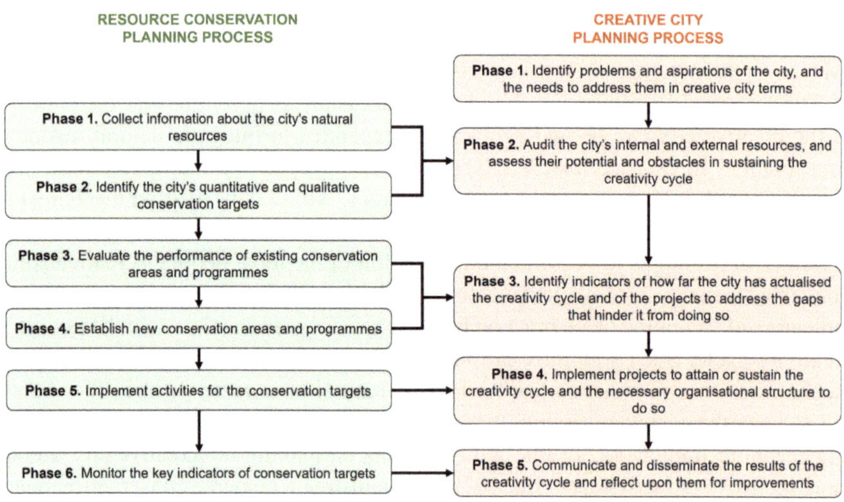

Fig. 9.3 Mainstreaming resource conservation planning in creative city planning

9.5 Discussion of Implications

The review of conservation strategies of the six UNESCO creative cities in the Philippines and other Asian countries through institutional analysis revealed that, apart from Iloilo (Philippines), Macao (China), and Singapore (a city-state), there is a notable lack of focus within creative city processes on addressing resource conservation challenges, despite the deep-rooted connection of local cultures to nature and the physical environment in general. To ensure the effective sustainability of the creative paradigm, continual attention to action situations is necessary to avoid missed opportunities for enhancing outcomes. This observation reflects the dominant Euro-American perspective inherent in the interpretation of the creative city, where natural resources are often undervalued, as argued by Alacovska and Gill (2019). Echoing Ostrom (2011), this setup has resulted in numerous missed opportunities for devising and implementing innovative solutions to enhance economic efficiency, fiscal equivalence, redistributional equity, accountability, conformity to local actors' values, and sustainability in the use of local natural resources within cities.

Conversely, the examples of Iloilo (Philippines), Macao (China), and Singapore (a city-state) underscore the significance of considering natural resources as integral to cultural resources within creative city processes, aligning with the core argument of this chapter. By elevating natural resources to the same level as cultural resources within creative city processes, this chapter proposes a reframing of the conceptual and planning trajectories for the establishment and growth of cities, emphasising the essential role of natural resources in both the creation and preservation of cultural assets (Eriksen, 2021). However, applying so may only present a formidable challenge for the local authorities of creative cities in developing countries like the Philippines that are already struggling to harmonise existing policies and programmes that could contribute in attaining sustainable development. Additionally, developed countries like the city-state of Singapore, whose cohesive planning generally results in positive socioeconomic outcomes, may struggle to recognise divergent cultural activities emerging from people's spontaneous interactions with nature. Nevertheless, these frameworks challenge and stress the need to review the existing interpretations and frameworks on the creative city, including those of the Philippine Creative Industries Development Act (2022), the UCCN (2023), and even European Union's Cultural and Creative Cities Monitor (Alberti et al., 2023).

9.6 Future Directions

In the case study, it is shown that while a strong public-private collaboration is a crucial step to develop a local creative economy, it falls short in sustaining creative city initiatives over the long term. To ensure long-term sustainability, there is a pressing need for the establishment and implementation of a cohesive local conservation policy that delineates a city's designation as a creative city. In the case of creative cities in the Philippines, the absence of this unified local policy has posed significant challenges in identifying their operational rules for resource conservation within the framework of creative city endeavours. However, in the case of creative cities in other Asian countries, Singapore (a city-state) has shown that even having a unified local policy may still pose challenges for highly regulated governments, particularly in recognising divergent and unplanned outcomes, such as organic cultural activities on the ground. This implies the need for a balanced and open interpretation of such a policy to ensure sustainable outcomes. Nevertheless, an examination of available data sources, including scholarly research and policy documentation, suggests that creative city development holds promise as a platform for promoting resource conservation. However, this finding is based on secondary data and is not conclusive, implying that primary research must be implemented to test the relationship in the field.

Given that this chapter exclusively examines cases from the Philippines and selected Asian countries, it becomes crucial to evaluate the broader applicability of the proposed conceptual and planning frameworks. This entails assessing their relevance in diverse settings, especially countries grappling with similar relationships with their natural resources. Ultimately, this chapter advocates for the integration of the creative city model as a strategic tool for advancing cities' efforts in natural resource conservation. This entails elevating the status of natural resources to be on par with cultural assets within the creative city framework. Moreover, it calls for a reassessment of the interpretations of creative city development processes by institutions responsible for setting standards, such as the UCCN. Additionally, reforms to existing national and global policies governing the development of creative cities are deemed necessary to align with these considerations.

9.7 Conclusion

The creative city has become an important model for the development of cities around the world, but its emphasis on the economic rewards of creativity undervalues and puts other resources of a city, such as natural resources, at risk of unsustainable outcomes. This chapter has shown that natural resources are as important as human creativity in ensuring the sustainable development of creative cities. The reviewed literature established the relationship between creative city making and resource conservation by looking at the necessity of natural resources in the creation and maintenance of cultural resources. This relationship is supported by the findings of the institutional analysis of the resource conservation processes of Philippine creative cities and their close Asian counterparts, wherein their general lack of attention to resource conservation challenges endangers the very cultural resources that define their designation as UNESCO creative cities. Consequently, this chapter attempted to integrate this relationship into the conceptual and planning frameworks of creative city making, expanding its scope as an urban governance tool. The proposed frameworks can be utilised in the review of the existing interpretations and standards for the creative city. However, more research is needed to effectively recalibrate the creative city framework and balance the tensions between its economic promises and its resource conservation needs.

Highlights

- The economic orientation of the creative city model jeopardises the natural resources of Asian creative cities.
- Natural resources are integral in sustaining cultural resources necessary for nurturing human creativity.
- Recognising natural resources as equally important as human creativity is essential for the future of creative city frameworks.

References

Alacovska, A., & Gill, R. (2019). De-westernising creative labour studies: The informality of creative work from an ex-centric perspective. *International Journal of Cultural Studies, 22*(2), 195–212. https://doi.org/10.1177/1367877918821231

Alberti, V., Caperna, G., Mauri, C., Panella, F., Tacao Moura, C. J., Banys, K., Saisana, M., et al. (2023). *Cultural and creative cities monitor.* Publications Office of the European Union. https://doi.org/10.2760/975348

An Act Providing for the Development and Promotion of the Philippine Creative Industries. (2022). *Republic Act No. 11904.* Official Gazette of the Philippines. Retrieved from https://lawphil.net/statutes/repacts/ra2022/ra_11904_2022.html

AuthentiCity. (2008). *Creative city planning framework.* Toronto Arts Council. Retrieved from https://torontoartscouncil.org/TAC/media/tac/Reports%20and%20Resources/City%20of%20Toronto/creative-city-planning-framework-feb08.pdf

Bettencourt, L. M., & West, G. B. (2017). *Cities become more creative and efficient as they grow.* Retrieved from Scientific American: https://www.scientificamerican.com/article/cities-become-more-creative-and-efficient-as-they-grow/

Cabra, J. F., & Guerrero, C. D. (2022). Regional creativity: Cultural and socio-economic differences. *Journal of Creativity, 32,* 1–5. https://doi.org/10.1016/j.yjoc.2022.100022

Castillo, A. (2023). *City to hold public poll on "vision" for the city.* Retrieved from The Baguio Chronicle: https://thebaguiochronicle.com/top-story/city-to-hold-public-poll-on-vision-for-the-city/

Chaudhury, S., & Lundberg, A. (2018). Singapore as a creative city: Vignettes from the perspective of la Flâneuse Tropique. *Electronic Journal of Studies in the Tropics, 17*(2), 110–131. https://doi.org/10.25120/etropic.17.2.2018.3660

Chuangchai, P. (2020). *The creative city in the Asean context: The cultural policy of the Southeast Asian creative cities network.* University of Warwick, Centre for Cultural and Media Policy Studies. Unpublished. Retrieved from https://wrap.warwick.ac.uk/160140/

City Government of Iloilo. (2023). *2023–2028 Iloilo city comprehensive development plan.* City Government of Iloilo, City Planning and Development Office. Iloilo. Retrieved from https://iloilocity.gov.ph/main/wp-content/uploads/2023/05/CDP2023-2028_4-13_Final-Document.pdf

de Guzman, R. S. (2020). A profiling study of the creative industry artists of Baguio City, the first creative city in the Philippines. *Journal of Economics, Management and Agricultural Development, 6*(1), 33–56. Retrieved from https://jemad.cem.uplb.edu.ph/articles/a-profiling-study-of-the-creative-industry-artists-of-baguio-city-the-first-creative-city-in-the-philippines/
DesignSingapore Council. (2015). *Design 2025*. DesignSingapore Council. Retrieved from https://designsingapore.org/resources/design-2025/
Elizaga, T. D., & Chua, J. V. (2021). From Sugbu to 'Ceboom': A study on the historical development of the urban form of Cebu City. *Espasyo: Journal of Philippine Architecture and Allied Arts, 8*(2018–19).
Eriksen, T. H. (2021). The loss of diversity in the anthropocene biological and cultural dimensions. *Frontiers in Political Science, 3*, 1–10. https://doi.org/10.3389/fpos.2021.743610
Freilich, M. (1967). Ecology and culture: Environmental determinism and the ecological approach in anthropology. *Anthropological Quarterly, 40*(1), 26–48. https://doi.org/10.2307/3316830
Gehl, J. (2012). Cities for people. *Journal of Architectural and Planning Research 29*(2), 172–174.
Huang, X. (2022). The role of cultural and creative industries in special economic zones: Development recommendations for the Macao-Hengqin region. In P. Figueiredo, F. Leandro, and Y. Li (Eds.), *Handbook of research on special economic zones as regional development enablers* (pp. 234–250). IGI Global Scientific Publishing. https://doi.org/10.4018/978-1-7998-7619-9.ch012
IUCN. (1980). *World conservation strategy: Living resource conservation for sustainable development*. International Union for Conservation of Nature and Natural Resources (IUCN). Retrieved from https://portals.iucn.org/library/efiles/documents/wcs-004.pdf
Jackson, A. (2023). *Classifications and representations of Macanese cuisine: From domestic kitchen to Casino Hotel*. Goldsmiths, University of London, Institute for Creative and Cultural Entrepreneurship. Goldsmiths, University of London. Retrieved from https://research.gold.ac.uk/id/eprint/34016/1/ICCE_thesis_JacksonA_2023.pdf
Joos-Vandewalle, S., Wynberg, R., & Alexander, K. A. (2018). Dependencies on natural resources in transitioning urban centers of northern Botswana. *Ecosystem Services, 30*(C), 342–349. https://doi.org/10.1016/j.ecoser.2018.02.007
Keong, C. Y. (2015). Sustainable resource management and ecological conservation of mega-biodiversity: The Southeast Asian big-3 reality. *International Journal of Environmental Science and Development, 6*(11), 876–882. https://doi.org/10.7763/IJESD.2015.V6.715
Landry, C. (2008). *The creative city: A toolkit for urban innovators* (2nd ed.). Earthscan.
Landry, C., & Hyams, J. (2012). *The creative city index: Measuring the pulse of the City*. Comedia. Retrieved from https://unesdoc.unesco.org/ark:/48223/pf0000261892
Macao SAR Government Tourism Office. (2021). *Macao tourism industry development master plan review report*. Government Tourism Office. Macao Special Administrative Region. Retrieved from https://masterplan.macaotourism.gov.mo/2021/download/Review_Report_Action_Plan_en.pdf
Macao Special Administrative Region. (2023). *Development plan for appropriate economic diversification of the Macao special administrative region (2024–2028)*. Macao Special Administrative Region. Retrieved from https://www.dsepdr.gov.mo/uploads/attachment/2024-03/e-book_E.pdf
Margules, C. R., & Pressey, R. L. (2000). Systematic conservation planning. *Nature, 405*, 243–253. https://doi.org/10.1038/35012251
Marrelli, M., & Fiorentino, P. (2016). Cultural commons and local art markets: Zero-miles contemporary art in Naples. *City, Culture and Society, 7*(2), 117–122. https://doi.org/10.1016/j.ccs.2015.12.002
Molho, J., Levitt, P., Dines, N., & Triandafyllidou, A. (2020). Cultural policies in cities of the "global south": A multi-scalar approach. *International Journal of Cultural Policy, 26*(6), 711–721. https://doi.org/10.1080/10286632.2020.1811256
Montera, D. E., & Lin, H. W. (2018). Peripheral realities: Visual artists in Cebu, Philippines. *Journal of Literature and Art Studies, 8*(9), 1357–1363. https://doi.org/10.17265/2159-5836/2018.09.007

O'Connor, J., & Shaw, K. (2014). What next for the creative city? *City, Culture and Society, 5*(3), 165–170. https://doi.org/10.1016/j.ccs.2014.05.010

Oabel, P. V. (2003). *The Iloilo urban region: Structural change, decentralization and employment in a Philippine secondary city.* MA Thesis, University of British Columbia, Department of Geography, Vancouver. Retrieved from https://open.library.ubc.ca/soa/cIRcle/collections/ubctheses/831/items/1.0091046

Ostrom, E. (2011). Background on the institutional analysis and development framework. *The Policy Studies Journal, 39*(1), 7–27. https://doi.org/10.1111/j.1541-0072.2010.00394.x

Polski, M. M., & Ostrom, E. (2017). An institutional framework for policy analysis and development. In D. H. Cole, and M. D. McGinnis (Eds.), *Elinor Ostrom and the Bloomington School of political economy* (Vols. 3, pp. 13–47). A framework for policy analysis. Lexington Books. Retrieved from https://ostromworkshop.indiana.edu/doc/teaching/iad-for-policy-applications.pdf

Pongruengkiat, W., Tippayawong, K. Y., Aggarangsi, P., Pichayapan, P., Katongtung, T., & Tippayawong, N. (2023). Assessing sustainability of Chiang Mai urban development. *Discover Sustainability, 4*, 1–22. https://doi.org/10.1007/s43621-023-00174-2

Pratt, A. C. (2010). Creative cities: Tensions within and between social, cultural and economic development—a critical reading of the UK experience. *City, Culture and Society, 1*, 13–20. https://doi.org/10.1016/j.ccs.2010.04.001

Rinchumphu, D., Srivanit, M., Iamchuen, N., & Aryupong, C. (2024). Exploring summer variations of driving factors affecting land use zoning based on the surface urban heat island in Chiang Mai, Thailand. *International Journal of Geo-Information, 13*(7), 1–27. https://doi.org/10.3390/ijgi13070228

Salas, V. (2023). *How Iloilo became a UNESCO creative city for gastronomy.* Retrieved from Positively Filipino https://www.positivelyfilipino.com/magazine/how-iloilo-became-a-unesco-creative-city-for-gastronomy

Santos, J. P. (2020). 'Crisis of values' in rhetorical architecture: Creative cities discourse and the urban landscapes of the Philippines. In X. Gu, M. K. Lim, & J. O'Connor (Eds.), *Re-imagining creative cities in twenty-first century Asia* (pp. 77–93). Springer Nature. https://doi.org/10.1007/978-3-030-46291-8

Segovia, C., & Hervé, J. (2022). The creative city approach: origins, construction and prospects in a scenario of transition. *City, Territory and Architecture, 9*(1), 1–15. https://doi.org/10.1186/s40410-022-00178-x

UCCN. (2023). *2023 Call for applications.* Retrieved from UNESCO Creative Cities Network https://en.unesco.org/creative-cities/content/2023-call-applications

UN Environment Programme. (1999). *Cultural and spiritual values of biodiversity.* In D. A. Posey (Ed.) Intermediate Technology Publications.

UN Habitat. (2020). *What is a city?* Retrieved from UN Habitat https://unhabitat.org/sites/default/files/2020/06/city_definition_what_is_a_city.pdf

World Bank Group. (2017). *Philippines urbanization review: Fostering competitive, sustainable and inclusive cities.* World Bank Group. Retrieved from https://www.worldbank.org/en/country/philippines/publication/fostering-competitive-sustainable-and-inclusive-cities-in-the-philippines

Wu, S., Lu, Y., & Fang, H. (2022). Evolution process of land reclamation in Macao and its impact on economy and ecology. In *Proceedings of the 2022 7th International Conference on Financial Innovation and Economic Development (ICFIED 2022)* (pp. 3067–3076). Atlantis Press International B.V. https://doi.org/10.2991/aebmr.k.220307.501

Jofer B. Asilum is an environmental planner who has worked with several local government units in the Philippines. He has a master's degree in Natural Resources Management from the Asian Institute of Technology in Thailand. His research areas include natural resources governance, rural political economy, asset formation, rural planning, and culture and arts planning.

Dayanara Bianca S. Bermudez is an environmental planner and GIS specialist. She has a postgraduate diploma in Urban and Regional Planning from the University of the Philippines Diliman. Her research interests include environmental sustainability, community resilience, and urban development strategies aimed at fostering equitable and resilient cities.

Chapter 10
From Cubicles to Connected Cities: The Future of Workspaces in Selected Cities in China, South Korea, Thailand, Malaysia, and the United Arab Emirates

Clusters of Future Studies: Environmental Futures; Technological Trends

Liyana Hasnan, Nurdiyana Zainal Abidin, and Noor Hashimah Hashim Lim

Abstract The work environment has transitioned from the traditional, static office configuration to a more adaptable model, spurred by the Fourth Industrial Revolution. The traditional cubicle-based workspace is no longer adequate to address the profound changes, and this shift in workspaces is intricately connected to transformative alterations in urban landscapes, initiating an era that redefines urban environments. The rise of remote work further propels the decentralisation of work, poised to reform cities with an emphasis on suburban resurgence and the development of mixed-use structures. The objective is to examine how evolving workspaces influence urban development and vice versa, with a focus on sustainability, flexibility, and technological integration. The chapter employs a conceptual methodology grounded in a qualitative research tradition to explore the future of workspaces to showcase their integration of sustainable practices, flexible design principles, and community engagement initiatives. These case studies illustrate how workspaces are shaping the future of urban environments in selected cities in China, South Korea, Thailand, Malaysia and the United Arab Emirates, fostering creativity, productivity, and collaboration while promoting environmental conservation and social inclusion. Here, the

L. Hasnan (✉) · N. Z. Abidin
Department of Architecture, Faculty of Built Environment, Universiti Malaya, Kuala Lumpur, Malaysia
e-mail: liyana_h@um.edu.my

N. Z. Abidin
e-mail: nurdiyana@um.edu.my

N. H. H. Lim
Department of Urban and Regional Planning, Faculty of Built Environment, Universiti Malaya, Kuala Lumpur, Malaysia
e-mail: shimalim@um.edu.my

R. Zainol et al. (eds.), *Future of Cities in Asia*,
https://doi.org/10.1007/978-3-032-01016-2_10

future of workspaces assumes a central role in shaping the spatial, technological, and cultural dynamics of cities, necessitating a comprehensive approach to urban planning that embraces flexibility, sustainability, and technological integration. The conclusion highlights the importance of embracing sustainable urban planning principles and flexible workspace designs to create vibrant, resilient, and inclusive cities for the future.

Keywords 4IR · Cities · Coworking · Future office · Workspaces · Urban environment

10.1 Introduction

Jacques Tati's iconic film "Play Time," released in 1967, offered a satirical glimpse into the impersonal and labyrinthine office environments of the modern era, characterised by glass and steel buildings and endless rows of cubicles. However, the days of these rigid office setups, reminiscent of scenes from the "Dilbert" comic strip or the television sitcom "The Office," are long gone. Today, the global workplace landscape is undergoing a seismic shift, catalysed by the COVID-19 pandemic. As organisations grapple with balancing remote work with returning to physical office spaces, there is a pressing need to rethink and reimagine how we manage and design office environments to accommodate the evolving workplace culture.

Prior to the 2020 pandemic, the work environment had already experienced a significant transformation in response to the Fourth Industrial Revolution (4IR). The 4IR, characterised by advancements in artificial intelligence, robotics, and automation, is transforming industries and job roles. This shift is characterised by the emergence of adaptable workspaces prioritising flexibility and innovation. Furthermore, advancements in workspace, technology, robotics, and artificial intelligence suggest new work designs (Schultz, 2021).

The 4IR enables flexible work arrangements, such as remote work and telecommuting, facilitated by digital connectivity and collaboration tools. It also facilitates the rise of the gig economy and entrepreneurship in Asia, as digital platforms enable individuals to offer services and create businesses independently. This trend has implications for work and urban environments, as it blurs the boundaries between formal and informal economies and fosters innovation and competition. In the context of future workspaces, the integration of technologies such as augmented reality and virtual reality is crucial. Dataspace, a reconfigurable hybrid reality environment, has been proposed as a tool for collaborative information analysis in future workspaces (Cavallo et al., 2019).

The development of autonomous mobile manipulators, a robot system built from a robotic manipulator arm mounted on a mobile platform for flexible production, reflects the integration of advanced technologies to support adaptable work environments (Engemann, 2020). Additionally, incorporating flexible workspaces within corporate real estate emphasises the growing recognition of the need for adaptable physical work environments to accommodate evolving work practices and preferences. Moreover, the study by Hancke and Silva (2012) highlights the significant role of advanced sensing technologies in smart cities, enabling improvements in various aspects such as building design and infrastructure monitoring.

Research has shown that workspace satisfaction is associated with self-satisfaction, good self-perceived future workability, and good recovery (Lusa et al., 2019). Additionally, the use of virtual and hybrid workspaces has been highlighted to improve job-site safety and enhance industry productivity (Santhi & Selvanayagam, 2025). Future workspaces also rely on physical elements, such as the presence of plants, non-crowded workspace, and window views, to stimulate creativity (Dul & Ceylan, 2014). Providing employees with sufficient workspaces is essential to organisational innovativeness.

The 4IR transition extends beyond the confines of individual workplaces. It intertwines with broader urban development trends, shaping the fabric of our cities. As companies gravitate towards more dynamic workspaces, urban landscapes undergo a metamorphosis, accommodating the needs of a workforce that values mobility and connectivity. The future of workspaces is a critical factor in shaping cities' spatial, technological, and cultural dynamics, necessitating a comprehensive approach to urban planning that embraces flexibility, sustainability, and technological integration.

Adaptable workspaces within the company emphasise physical layout and internal usage, raising issues such as non-territorial spaces, hot-desking and hotelling (Silva, 2021). Technology has allowed the office space to evolve beyond its industrial heritage; space as a service is an opportunity for corporations to explore flexibility. Technology has created an era where boundaries between work and life blur, and the traditional nine-to-five paradigm has become increasingly obsolete. It challenges the conventional notions of productivity and efficiency, emphasising the importance of holistic wellbeing and work-life balance.

Urban planning is needed to accommodate flexible workspaces in the future. Nevertheless, amidst this wave of change, questions arise about the implications for social dynamics, equity, and sustainability. Will these new workspaces foster inclusivity and diversity, or will they exacerbate existing inequalities? Evolving workspaces have become powerful in influencing productivity, employee satisfaction, and innovation. It is critical now that the workplace responds not just to different industries and job functions but also to organisational and individual needs.

The evolution of the workspace globally has moved from rigid, hierarchical office environments dominated by closed cubicles and centralised control to more fluid, decentralised, and technology-enhanced models. These changes have been driven by technological advancements, shifts in organisational culture, and, more recently, by the pandemic-induced acceleration of remote and hybrid work. Across Asia, this transformation is further shaped by rapid urbanisation, diverse economic development trajectories, and strong governmental influence on infrastructural planning.

The selection of China, South Korea, Thailand, Malaysia, and the United Arab Emirates (UAE) reflects a strategic focus on urban centres with varying technological readiness, socio-cultural norms, and economic structures. Each of these regions demonstrates a distinct model of adapting to the future of work: China with its smart city ecosystems; South Korea through high digital penetration and corporate innovation; Thailand and Malaysia via emerging knowledge economies and regional planning experiments; and the UAE as a global hub integrating futuristic urbanism with economic diversification.

The evolution of the modern work environment is not merely a superficial change in office aesthetics but a profound response to the transformative forces reshaping the professional landscape. The future of workspaces requires a multidimensional approach that integrates employee satisfaction, technological advancements, analytical skills, and flexible workspace design within urban planning. It is imperative to critically scrutinise the design and implementation of new work environments. By considering these factors, urban planners and organisations can create workspaces that promote well-being, productivity, sustainability, and technological integration towards a more adaptable and connected future. This chapter highlights the current practice of workplaces in Asia, especially China, South Korea, Malaysia, Thailand, and the UAE, and how physical workspaces are challenged and redesigned. It also reveals a glimpse of future workspaces in the region.

This research aims to examine how the convergence of technological innovation, public health considerations, and shifting work patterns is influencing the form, function, and future of urban workspaces. Conceptually, the paper frames this inquiry through the lens of socio-technical systems and generative design thinking, proposing that future workspaces should not only respond to contemporary demands but also proactively shape urban possibilities through design-led innovation.

10.2 Methodology

This paper adopts a conceptual methodology grounded in a qualitative research tradition to explore the future of workspaces within the evolving urban contexts of selected Asian cities. As a concept research, its primary aim is to critically examines the reciprocal and generative relationship between post-industrial workspace transformations and urban spatial restructuring.

To anchor these theoretical propositions in real-world conditions, the paper employs a comparative case study strategy. The selected cities—Shanghai, Seoul, Bangkok, Kuala Lumpur, and Dubai—serve as illustrative examples to support the conceptual arguments. These cities are chosen for their strategic positioning as regional economic hubs, active digital transformation participation, visible spatial responses to technological shifts, and the post-pandemic imperative.

Examples of workspaces from each cities were analysed, drawing from urban planning, smart city frameworks, Architectural and workspace design initiatives, and post-COVID urban adaptation. It explores how digital technologies, organisational culture, and urban development trends intersect to reshape how and where people work.

This method allows the research to generate insights inductively, using existing urban practices to provoke new theoretical imaginaries about the future of work and cities. The cases are not meant to represent exhaustive empirical evidence but to illustrate, challenge, and extend the conceptual framing. In doing so, the methodology maintains a generative and speculative stance, aligning to contribute to scholarly discourse and the design of adaptive, future-oriented built environments.

The research is contextualised within the Asian region, highlighting how cultural, economic, and policy differences affect adopting these workspace trends. Through this multi-layered approach, the paper aims to present a clear picture of how future workspaces are taking shape and what this means for urban planning, technology integration, and the future of work.

10.3 Key Concepts and Issues

The evolution of workspaces across the selected cities—China, South Korea, Thailand, Malaysia, and the UAE—illustrates a dynamic shift from traditional cubicle environments towards more integrated, flexible, and technology-driven spaces. This transition reflects broader economic trends, cultural shifts, and technological innovations influencing how people work and interact within urban frameworks.

1. Flexible Workspaces and Employee Satisfaction

A critical concept in the evolution of workspaces is the emphasis on flexibility and the holistic well-being of employees. Research indicates that organisations have increasingly recognised the necessity for work environments that foster creativity and employee happiness. For instance, companies are now designing modern workspaces beyond conventional office layouts, accommodating diverse work habits such as remote working and collaborative environments (Kaur & Solomon, 2022; Margariti et al., 2021). The shift emphasises creating spaces that support happiness and productivity, aligning with the need for employee-centric designs that facilitate multifunctionality, such as integrating areas for collaboration alongside personal workspaces (Pakos et al., 2023).

2. Technological Integration and Workspace Optimisation

Integrating disruptive technologies plays a pivotal role in reshaping work environments in these cities. In Thailand, for instance, adopting advanced technologies—including artificial intelligence—has revolutionised business operations, particularly in human resource recruitment practices, which are becoming more automated and efficient (Wongras & Tanantong, 2023; Tanantong & Wongras, 2024). In the urban context of Dubai, embracing smart city initiatives enhances the functionality of workspaces, promoting seamless connectivity and interaction among users (Ma et al., 2024; Amrousi et al., 2019). Moreover, these technologies are also applied to benchmark workspace usage, optimising physical environments to suit modern work patterns across various sectors (Kaur & Solomon, 2022).

3. Cultural Influences and Work Dynamics

Cultural factors significantly influence how workspace evolution is perceived and implemented within different countries. In South Korea, a traditionally hierarchical work culture slowly adapts to more egalitarian practices that encourage work-life balance and flexibility. This is essential as issues like long working hours and job dissatisfaction remain prevalent (Yoon and Park 2023; Hite and McDonald 2020). Alternatively, in Malaysia, the push towards developing a knowledge-based economy through innovation highlights the importance of modern workspaces that cater to young professionals looking for engaging environments that stimulate creativity and collaboration (Chairaksa & Pankham, 2023).

4. Urban Planning and Sustainable Development

Urban planning intertwines with workspace evolution, especially in cities like Dubai and Bangkok. Both cities face challenges in connectivity and sustainability, necessitating thoughtful urban design that prioritises accessible workspaces integrated within broader urban settings (Alawadi & Dooling, 2015; Zabanoot and Bleibleh 2019). For instance, Dubai's urban development involves creating multifunctional public spaces that promote work, recreation, and social interaction, helping foster a connected urban community (Amrousi et al., 2019; Zabanoot & Bleibleh, 2019). Sustainability becomes a focal point, as cities like Bangkok strive to address environmental impacts while enhancing the quality of urban work environments (Thushari et al., 2020).

5. Future Directions for Workspaces

Looking forward, the evolution of workspaces in these diverse cities underscores a need for agility in workspace design, reflecting the rapidly changing expectations of workers in a post-pandemic era. Emphasising mental well-being, promoting hybrid work models, and integrating advanced technologies will be paramount in shaping the future of workspaces. Moreover, understanding the socio-economic impacts of these changes is essential as urban centres continue to evolve in response to global trends and local cultural nuances (Warunyu et al., 2024).

Technological advancements, cultural shifts, and urban development strategies drive the transition from traditional cubicle setups to more innovative, interconnected work environments. Adapting to these dynamics will be critical for fostering effective and sustainable workspaces in China, South Korea, Thailand, Malaysia, and the UAE.

10.3.1 Workspaces and Urban Environments in Asian Countries

In Asia, 4IR has significantly impacted the economy, particularly in the manufacturing and services sectors. Asian countries are transitioning from traditional economies to knowledge-based ones driven by services, technology, and innovation, leading to the emergence of tech hubs, business districts, and creative clusters.

Asian cities are becoming global innovation centres, attracting skilled labour and investments. Research by Hartley et al. (2018) emphasises the role of process innovation in technology, particularly in Asian cities like Seoul, Singapore, Taiwan, and Hong Kong, in fostering economic clustering. These changes align with the necessity for increased concentration and complex economic densities in cities like Shanghai to compete with global urban centres such as London, New York, Tokyo, Seoul, Singapore, and Hong Kong (Salat, 2017).

The convergence of technological advancements, demographic changes, and diverse cultures in this region presents the region with a unique strategy to brand itself as a global hub, reflecting the evolving nature of workspaces in these regions (Joo & Seo, 2017). Traditional hierarchical office structures are giving way to more open and collaborative workspaces to respond to the rise of start-ups, technology firms, and coworking spaces, reshaping the fabric of urban centres (Anttiroiko, 2014; Ishitsuka, 2020).

Not only new modes of work but also other factors drive this transition, including the need to adapt to the changing dynamics of work environments, the desire to enhance communication and collaboration among employees, and the rising cost of real estate (Weziak-Bialowolska et al., 2018). The relationship between workspaces and urban environments in Asia extends beyond the confines of office buildings. It intersects with broader urban development initiatives to address congestion, pollution, and quality of life issues. Remote work, telecommuting, and coworking spaces are also changing the way people commute and interact in urban environments. At a time when Asian cities are facing issues of congestion and pollution, embracing flexible work practices can reduce the reliance on office commutes and enhance work-life balance.

The rise of the gig economy and digital platforms has enabled individuals to offer services and create businesses independently. This trend has implications for both work and urban environments, as it blurs the boundaries between formal and informal economies and fosters innovation and competition. Cities across the region are experimenting with mixed-use developments, transit-oriented development, and green spaces to create more liveable and sustainable urban environments.

Nonetheless, rapid urbanisation, socioeconomic disparities, and environmental degradation pose challenges; informal workspaces in the informal economy remain largely unaddressed. The dynamic landscape of Asia requires policymakers, urban planners, and businesses to collaborate to ensure that the evolution of workspaces contributes to inclusive and sustainable urban development. This entails investing in public transportation, affordable housing, and green infrastructure while promoting innovation and entrepreneurship.

10.3.2 Flexibility, Sustainability, and Technological Integration

Asian cities are harnessing technology to drive urban development, efficiency, and innovation. Through smart city initiatives, data analytics, Internet of Things (IoT) sensors, and digital platforms are employed to optimise urban services, enhance public safety, and elevate resident quality of life. From Shanghai and Guangzhou's smart transportation systems to Seoul's digital governance platforms, technological integration is revolutionising urban planning, management, and experiences.

Tencent's Future City in Shenzhen, China, epitomises visionary urban development, spanning two million square meters and prioritising pedestrians, green spaces, and self-driving vehicles. This forward-thinking project seamlessly integrates office spaces, residential areas, retail outlets, and green spaces into a sustainable urban environment (Holland, 2020). Offering company residences, offices, shops, schools, and public amenities connected by public transportation, Tencent's city fosters an interconnected human ecosystem. Smart technologies like IoT sensors, AI algorithms, and autonomous vehicles optimise resource utilisation, energy efficiency, and transportation systems. Flexible work arrangements and digital collaboration platforms enable seamless work across locations and time zones. Environmental sustainability is paramount, with rooftop solar panels and sophisticated wastewater capture and reuse systems embedded in the master plan.

Samsung Digital City in Suwon, South Korea, exemplifies the convergence of technology, sustainability, and wellbeing. Covering 390 acres, it offers flexible workspaces, free healthcare, and childcare facilities. The campus features recreational amenities like sports fields, a rock-climbing wall, and a theme park. Green initiatives include green roofs and solar panels. Advanced technologies such as AR displays enhance communication (Samsung Newsroom, 2014).

Alibaba Hangzhou Xixi Campus, in China, showcases innovative design and technology integration. It offers flexible work areas, AI-powered assistants, and wellness amenities. The campus aims to balance operational efficiency with employee satisfaction and is committed to being carbon–neutral by 2030. Self-service lobbies with AI reception and connected equipment enhance convenience. Another example is Toyota's Woven City at the base of Mount Fuji, which is designed to be a testing ground for technologies like robotics, smart homes, and artificial intelligence. Residents expected to move into the Woven City will live in smart homes with in-home robotics systems to assist with daily living and sensor-based artificial intelligence to monitor health and other basic needs.

Cyberport in Hong Kong, China is a digital community and incubator hub for start-ups, tech companies, and creative professionals. Spanning 25 hectares, it offers state-of-the-art facilities, including coworking spaces and maker labs. Technology integration, such as high-speed internet and smart building systems, fosters connectivity and efficiency. Collaborative initiatives like hackathons drive innovation and knowledge sharing.

Recognising the importance of community engagement and social inclusion in urban planning, Asian cities are adopting participatory approaches to their workspaces. By soliciting input from diverse stakeholders, cities can ensure that development projects reflect the needs, preferences, and aspirations of local communities.

Common Ground KL Eco City in Kuala Lumpur, Malaysia, is a coworking space within the sustainable KL Eco City development. It offers diverse work environments, wellness rooms, nap pods, and gaming areas. Regular networking events and skill-building sessions create a collaborative community of entrepreneurs and professionals. Impact Hub Dubai is a coworking and innovation space in Dubai that caters to social enterprises and sustainable businesses. It provides private offices, hot desks, innovation labs, and maker spaces. Cutting-edge technology tools empower members to drive positive social impact. Social impact events and sustainability workshops cultivate an inclusive community of changemakers.

Sustainability is now a cornerstone of urban planning in Asia, driven by concerns over climate change and environmental degradation. Cities like Singapore and Taipei are investing in green infrastructure and sustainable transportation to reduce carbon emissions and enhance resilience to natural disasters. For instance, green roofs and rainwater harvesting systems mitigate the urban heat island effect while promoting biodiversity.

The Hive Thonglor in Bangkok, Thailand, exemplifies this trend with its focus on sustainability and flexible design. This coworking space offers various work settings, from hot desks to private offices, while prioritising eco-friendly materials and energy-efficient systems. Biophilic design elements like green walls create a vibrant atmosphere, fostering creativity and productivity. The Hive Thonglor hosts events promoting wellness and education, encouraging collaboration and self-improvement.

Similarly, Punspace Nimman in Chiang Mai, Thailand, embraces sustainability and community engagement. This coworking space provides dedicated desks, communal areas, relaxation corners, and outdoor terraces for networking. By incorporating energy-efficient lighting and eco-friendly materials, Punspace Nimman aligns with Chiang Mai's commitment to environmental conservation.

Table 10.1 (see the next page) presents the workspaces in cities across Asia, describing them through four (4) features: (1) innovative features, (2) flexibility, (3) technology integration, and (4) sustainability.

10.4 Discussion of Implications

The 4IR is characterised by advancements in artificial intelligence, robotics, and automation, which are transforming industries and job roles. This disruption is particularly pronounced in Asia, where manufacturing and service sectors significantly contribute to the economy. Workspace awareness is a key aspect that influences collaboration in work environments. It provided users with information about ongoing activities in the collaborative space, enabling them to contribute effectively

Table 10.1 Workspaces in selected Asian countries described by innovative features, flexibility, technology integration, and sustainability

Workspaces	Countries	Innovative features	Flexibility	Technology integration	Sustainability
Alibaba Hangzhou Xixi Campus	China	Communal areas for diverse work styles, collaboration pods	Company campus, promote work-life balance	AI-powered assistants, VR-enabled meeting rooms, smart lighting	Carbon neutral by 2030
The Hive Thonglor, Bangkok	Thailand	Variety of work environments, activity-based workspace	Co-working	High-speed internet	Community engagement with variety of events hosted, biophilic design elements, eco-friendly materials
Tencent's Future City, Shenzhen	China	Mixed-use development, integrates office and residential	Interconnected work and private life	IoT sensors, AI algorithms, autonomous vehicles and digital collaboration platform	Rooftop solar panels, waste water recycle system
Samsung Digital City, Suwon	South Korea	Mixed-use development, integrates office, residential, schools, sport fields, swimming pool and theme park	Employees choose when and where to work	Advanced technologies such as AR for communication	Community support, green roofs
Cyberport, Hong Kong	China	Mixed-use development—multimedia park, incubator hub and maker labs	Co-working	High-speed internet, smart building system	Community engagement with variety of events hosted
Punspace Nimman, Chiang Mai	Thailand	Communal areas for diverse work styles, meditation corners and outdoor terraces	Co-working with options for flexible workspace usage	High-speed internet and streamlined collaboration	Energy-efficient lighting, eco-friendly materials

(continued)

Table 10.1 (continued)

Workspaces	Countries	Innovative features	Flexibility	Technology integration	Sustainability
Common Ground KL Eco City, Kuala Lumpur	Malaysia	Themed collaboration zones and variety of work environments, wellness room, nap pods and gaming area for recreation	Co-working, access to a network of Common Ground location across Southeast Asia for convenience of remote work	High-speed internet, smart lockers	Community engagement with variety of events hosted
Impact Hub, Dubai	UAE	Variety of work environments, prototyping labs, maker spaces for hands-on experimentation, rooftop garden	Co-working, access to a network of Impact Hub location across the world	Technology tool such as 3D printers, VR equipment, and collaborative project platforms	Community engagement with variety of events hosted and accelerator programs

to achieve shared objectives (Santos-Torres, 2022). Open-plan offices have gained popularity over the years due to their potential to increase organisational collaboration and communication (Weziak-Bialowolska et al., 2018). These spaces provide a platform for employees to interact, share ideas, and work together (Castillo-Cuesta et al., 2022).

Nonetheless, the future of workspaces in Asia extends beyond traditional open-plan designs, incorporating innovative approaches prioritising flexibility, well-being, and collaboration. Below are some modes of what to expect in workspaces in the future.

i. Hybrid-Work Environments—The rise of remote work and flexible schedules has reshaped the traditional office concept, combining physical office spaces with remote work options. This method allows employees to choose their preferred mode of working. Future workspaces are blurring the lines between work and play.
ii. Activity-Based Workspaces—Activity-based environments cater to different tasks and workstyles, including but not limited to quiet zones for focused work, collaborative spaces for group meetings, zones for relaxation, and zones for play and socialisation. These spaces allow companies to optimise productivity and creativity while enhancing employee well-being.
iii. Modular Design and Flexible Furniture—flexibility is key in future workspaces, with furniture and layout designs that can easily adapt to changing needs and preferences. Modular furniture systems allow for easy reconfiguration of spaces, enabling companies to optimise space utilisation and accommodate evolving work requirements. This flexibility promotes creativity, fosters collaboration, and enhances the overall user experience.
iv. Biophilic Design and Wellness Features—recognising the importance of employee well-being, future workspaces in Asia will prioritise biophilic design principles and wellness features. Greenery, natural light, and outdoor spaces create a connection to nature, reduce stress, and enhance cognitive function. Wellness amenities such as fitness centres, meditation rooms, and healthy food options promote physical and mental health, leading to happier and more productive employees.
v. Technology integration—advanced technologies play a central role in shaping the workspaces of the future. IoT sensors and smart lighting systems to virtual reality (VR) collaboration tools and artificial intelligence (AI) powered assistant technology will enhance efficiency, connectivity, and innovation in the workplace. Digital platforms and cloud-based solutions will facilitate seamless communication and collaboration among distributed teams, regardless of geographic location.

Sustainable Practices will be integral to future workspaces in Asia, with a focus on energy efficiency, waste reduction, and eco-friendly materials. Green building certifications such as Leadership in Energy and Environmental Design (LEED) and Building Research Establishment Environmental Assessment Method (BREEAM), will become standard requirements, driving the adoption of energy-efficient HVAC systems, recycled materials, and renewable energy sources. Sustainable commuting options such as bike-sharing programs and electric vehicle charging stations will reduce environmental impact and promote sustainable lifestyles.

10.5 Future Directions

The evolution of workspaces is reshaping Asia's economic, social, and urban landscapes. As Asian economies transform into service and knowledge-driven models, adaptable workspaces, such as coworking spaces in key hubs like Singapore and Kuala Lumpur, will continue attracting talent and investments while fostering innovative ecosystems.

The future of work is inextricably linked to urban planning. The changes in working arrangements immediately shift the spatial distribution of office spaces, affecting transportation networks, housing markets, and commercial real estate.

Another implication of future workspaces is their role in addressing mental health challenges in fast-paced urban environments, making the demand for employee wellness programs, ergonomic design, and stress-reducing environments grow. The design of flexible, inclusive and accessible work environments has become crucial in promoting employee engagement, satisfaction and well-being.

10.6 Conclusion

As Asia continues to drive global economic activity, creating conducive work environments that foster creativity, collaboration, and productivity becomes paramount. Policymakers and businesses can leverage insights into workspace evolution to support entrepreneurship, attract talent, and stimulate innovation ecosystems. Furthermore, workspace transformation is intricately linked to urban development and infrastructure planning. Optimising workspaces can play a crucial role in shaping sustainable and liveable urban environments as Asian cities grapple with rapid urbanisation, congestion, and environmental challenges.

The future of workspaces in selected cities across China, South Korea, Thailand, Malaysia, and the United Arab Emirates is positioned at the intersection of technological advancement, evolving workplace dynamics, and collaborative environments. Research in this area can inform urban planners about the spatial and design requirements of modern work environments, guiding decisions related to zoning, transportation, and public amenities. The way people work and interact in modern workspaces also has profound implications for workforce dynamics and well-being. Understanding cultural norms, societal expectations, and individual preferences in the Asian context can inform human resource policies, organisational practices, and workplace interventions aimed at enhancing employee well-being and productivity.

Moreover, the evolution of workspaces intersects with issues of social equity and inclusion. By examining how different population segments engage with evolving work environments, researchers can identify barriers to inclusion and propose strategies to promote diversity, equity, and social cohesion in the workforce.

Lastly, embracing digital technologies and sustainable urban planning initiatives can contribute to creating vibrant, resilient, and inclusive cities in Asia. Addressing challenges such as job displacement, urban congestion, and digital divides requires proactive policies and collaborative efforts from governments, businesses, and civil society stakeholders. By harnessing the transformative potential of modern workspaces and urban environments, Asian countries can chart a path towards more resilient, equitable, and vibrant futures for all.

Highlights

- The future of workspaces in cities across China, South Korea, Thailand, Malaysia, and the United Arab Emirates (UAE) is poised for a comprehensive transformation driven by technology and collaborative environments.
- Workspaces are no longer just physical offices.
- Urban development and workspaces are now tightly linked.
- Technology is the backbone of future workspaces.
- Well-being, flexibility, and sustainability are keys to future workplaces.
- Asia is leading with innovative real-world examples.

References

Anttiroiko, A. (2014). City branding as a response to global intercity competition. *Growth and Change, 46*(2), 233–252. https://doi.org/10.1111/grow.12085

Castillo-Cuesta, L., Ochoa-Cueva, C., & Cabrera-Solano, P. (2022). Virtual workspaces for enhancing collaborative work in EFL learning: A case study in higher education. *International Journal of Emerging Technologies in Learning (IJET), 17*(02), 4–18. https://doi.org/10.3991/ijet.v17i02.25937

Cavallo, A., & Ghezzi, A., & Balocco, R. (2019). Entrepreneurial ecosystem research: present debates and future directions. *International Entrepreneurship and Management Journal, 15.* https://doi.org/10.1007/s11365-018-0526-3

Dul, J., & Ceylan, C. (2014). The impact of a creativity-supporting work environment on a firm's product innovation performance. *Journal of Product Innovation Management, 31*(6), 1254–1267. https://doi.org/10.1111/jpim.12149

Engemann, H., Du, S., Kallweit, S., Cönen, P., & Dawar, H. (2020). Omnivil—an autonomous mobile manipulator for flexible production. *Sensors, 20*(24), 7249. https://doi.org/10.3390/s20247249

Hancke, G. P., & Silva, B. (2012). The role of advanced sensing in smart cities. *Sensors, 13*(1), 393–425. https://doi.org/10.3390/s130100393

Hartley, K., Woo, J. J., & Chung, S. K. (2018). Urban innovation policy in the postdevelopmental era: Lessons from Singapore and Seoul. *Asia & The Pacific Policy Studies*, *5*(3), 599–614. https://doi.org/10.1002/app5.255

Holland, O. (2020). CNN. Tencent is building a Monaco-sized' city of the future' in Shenzhen. CNN Style. https://edition.cnn.com/style/article/tencent-shenzhen-net-city/index.html

Ishitsuka, A. (2020). Visual encounters in global shanghai: On the desirability of bodies in a coworking space. *China Perspectives*, *2020*(3), 39–46. https://doi.org/10.4000/chinaperspectives.10416

Joo, Y., & Seo, B. K. (2017). Transformative city branding for policy change: The case of Seoul's participatory branding. *Environment and Planning C: Politics and Space, 36*(2), 239–257. https://doi.org/10.1177/2399654417707526

Lusa, S., Käpykangas, S., Ansio, H., Houni, P., & Uitti, J. (2019). Employee satisfaction with working space and its association with well-being—a cross-sectional study in a multi-space office. *Frontiers in Public Health, 7*. https://doi.org/10.3389/fpubh.2019.00358

Ma, C., Somrak, T., Manajit, S., & Gao, C. (2024). Exploring the potential synergy between disruptive technology and historical/cultural heritage in thailand's tourism industry for achieving sustainable development in the future. *International Journal of Tourism Research*, *26*(5). https://doi.org/10.1002/jtr.2759

Margariti, E., Ali, R., Grave, R., Verweij, D., Smeddinck, J., & Kirk, D. (2021). Understanding the experiences of remote workers: opportunities for ambient workspaces at home. *Frontiers in Computer Science, 3*. https://doi.org/10.3389/fcomp.2021.673585

Pakos, O., Eismann, T., Meinel, M., & Voigt, K. (2023). "more than meets the eye": unveiling the full potential of creative workspaces in modern organizations. *European Journal of Workplace Innovation*, *7*(2), 49–77. https://doi.org/10.46364/ejwi.v7i2.1033

Samsung Newsroom. (2014). 20 Things you didn't know about Samsung's headquarters in Suwon. https://news.samsung.com/global/purposely-leaked-20-things-about-samsungs-digital-city

Salat, S. (2017). Shanghai in the new economy of global flows shaping urban form and connectedness in the digital age. *Journal of Architecture and Urbanism, 41*(2), 145–161. https://doi.org/10.3846/20297955.2017.1325582

Santhi, V. & Selvanayagam, A. (2025). The Role Of Hybrid Work Models In Enhancing Employee Well-Being, Productivity, And Job Satisfaction. *South Eastern European Journal of Public Health*. XXVI. 3049–3062

Santos-Torres, A. (2022). Comparing visual representations of collaborative map interfaces for immersive virtual environments. *IEEE Access, 10*, 55136–55150. https://doi.org/10.1109/access.2022.3176949

Schultz, C. (2021). *The future of HR*. https://doi.org/10.5772/intechopen.96672

Tanantong, T., & Wongras, P. (2024). A utaut-based framework for analyzing users' intention to adopt artificial intelligence in human resource recruitment: a case study of thailand. *Systems*, *12*(1), 28. https://doi.org/10.3390/systems12010028

Thushari, I., Vicheanteab, J., & Janjaroen, D. (2020). Material flow analysis and life cycle assessment of solid waste management in urban green areas, thailand. *Sustainable Environment Research*, *30*(1). https://doi.org/10.1186/s42834-020-00057-5

Warunyu, S., Raphassorn, K., Weena, N., Koblap, A., & Pinnapa, M. (2024). Role of advanced transportation management systems in enhancing the distribution of frozen food from thailand to japan. *British Journal of Management and Marketing Studies*, *7*(1), 47–68. https://doi.org/10.52589/bjmms-7bp91qgb

Weziak-Bialowolska, D., Dong, Z., & McNeely, E. (2018). Turning the mirror on the architects: A study of the open-plan office and work behaviors at an architectural company. *Frontiers in Psychology, 9*. https://doi.org/10.3389/fpsyg.2018.02178

Liyana Hasnan is a researcher and Senior Lecturer at Universiti Malaya whose work investigates the cultural, historical, and political rhetoric within the discourses that have defined the architecture discipline. She earned her doctorate from Universiti Putra Malaysia. Her recent research looks at architectural pedagogy modes, spatial politics, and architecture as an interface. Between 2013 and 2015, she embarked on her master's degree at Columbia University, GSAPP, in the programme Critical, Curatorial, and Conceptual Practises in Architecture (MSc. CCCP).

Nurdiyana Zainal Abidin is a Senior Lecturer at Universiti Malaya, specialising in heritage and conservation within the field of architecture. She earned her PhD in Architecture from Universiti Teknologi Malaysia (UTM), focusing on sustainable vernacular architecture. Recently, her research has delved into a comparative analysis between traditional machiya in Kyoto and the traditional townhouse in Malacca, as well as inclusive educational empowerment. Additionally, she has broadened her research scope to investigate the intersection of heritage, archives, and digital technology.

Noor Hashimah Hashim Lim is a Senior Lecturer in Urban and Regional Planning at Faculty of Built Environment, Universiti Malaya, Malaysia. She obtained her PhD in Urban and Regional Planning from Universiti Teknologi Malaysia and MSc Project Management from the National University of Singapore. Her research interest revolves around statistical modelling, examining the intricate interplay between environmental factors and human behaviours, with a keen emphasis on population health, climate-community resilience, participatory planning and digital urbanism. Her recent works include developing a virtual community mapping that captures people–place attachment and engaging underprivileged B40 children in participatory design initiatives. Before becoming an academic, Hashimah had 7 years of industrial work experience in project management focusing on property development. She has also served as the Head of Research at a local think-tank in Malaysia, producing policy briefs related to urban and rural governance, city planning and urbanisation issues.

Index

R. Zainol et al. (eds.), *Future of Cities in Asia*,
https://doi.org/10.1007/978-3-032-01016-2

MIX
Papier aus verantwortungsvollen Quellen
Paper from responsible sources
FSC® C105338

If you have any concerns about our products,
you can contact us on
ProductSafety@springernature.com

In case Publisher is established outside the EU,
the EU authorized representative is:
Springer Nature Customer Service Center GmbH
Europaplatz 3, 69115 Heidelberg, Germany

Printed by Libri Plureos GmbH
in Hamburg, Germany